What's Inside the Black Box?

A look at modern technology and
how some of the familiar things
around us actually work

By

Doug Domke

Table of Contents

Chapter 14 - Computer Memory   97

If you use a computer, have a cell phone or digital camera, you've heard all the acronyms: RAM, ROM, Flash memory, and hard drive. Why do we have all these different kinds of computer memory? What do they all do? And for that matter, what does any computer memory actually do?

Chapter 15 – Solar Panels   103

This is a continuation of our discussion of commercial electric power. We'll look at how solar panels work and how they fit into the commercial power scheme.

Chapter 16 – A Computer's CPU   109

Talk about a high-tech black box. Your computer's central processing unit is a mystery to even most computer programmers. We'll explain what goes on inside the CPU.

Chapter 17 – Digital Cameras   117

Digital cameras have almost completely replaced conventional cameras today. But how do they differ from traditional film cameras? And how do they work?

Chapter 18 – The Aircraft Black Box   123

With a whole book about black boxes, we can't ignore the aircraft black box – what does it do and how does it work?

Final Thoughts   127

Preface

This book resulted from combining two ideas I had for a second book.

The first idea is about how modern technology is purposely compartmentalized to allow large groups of people to work collectively on very complex products. If the inputs and outputs of a given system or technology are defined well enough, others can use it without any knowledge of what's inside or how the internal parts work. This is the black box concept, which is discussed at length in Chapter 1.

The second idea is about explaining in easily understandable terms how complex pieces of technology that we use every day actually work. We use these things easily and effortless without knowing what's inside or how they do what they do. They are black boxes to us. So in the remaining chapters, we take some of these items apart and see how they work.

These two ideas are combined in this book to talk about "What's Inside the Black Box?" I hope you enjoy!

The chapters can be read in any order. But many of the technologies we talk about in this book involve electricity and electronics. So in Chapter 5 on Commercial Electric Power, I have included a brief tutorial on basics of electricity. It may be helpful to read this before reading the later chapters, if you have limited knowledge of electricity.

Finally, I would like to thank my wife Polly for her help with editing and proofreading.

Doug Domke
May 2014

Chapter 1 – The Black Box Concept

When you hear the term "black box", perhaps the first thing that comes to mind is the black box on airplanes – the cockpit voice recorder and flight data recorder that are aboard all commercial aircraft today. We are actually going to talk about aircraft black boxes in our last chapter, but the meaning of "black box" in this book is more generic. It's a concept, rather than a physical black box. And it's all about technology and how it evolves. It's about how very complex projects are broken down into manageable pieces.

Technology is all around us today, and for most of us, the inner workings are incomprehensible. What you might not know is that most of these inner workings are incomprehensible even to the people who design them. That's because modern technology is the result of collective intelligence, where one person may make a significant contribution, but doesn't know and doesn't need to know how the complete system works. We get great technology by compartmentalizing complex technological challenges. One guy figures out how a certain combination of inputs leads to and generates a specific output or combination of outputs. How he does it isn't important to the rest of his team. What is important is that he documents all the inputs and all the outputs. Once that's done, everyone else on his team can use it. The inner workings can remain unknown to everyone else. It's just a "black box" to them, but it does something they need.

This concept of the "black box" can also be called "encapsulation". It means we wrap up a nice piece of working technology into a documented package, with meticulously defined inputs and outputs, instructions on how to use it, and then we basically hide the internal workings. We hide them because we don't want everyone else to have to worry about how the insides operate. We just want our packaged piece of technology to be available for others to use.   The black box notion of encapsulation permeates modern technology in all fields large and small. It's what allows us to break up incredibly complicated tasks and technologies into small pieces that an individual can conquer. By partitioning a complex technology into smaller, more manageable projects or tasks, it can be distributed across an organization. Individuals work on various pieces, while others work on integrating those pieces into the finished product. The result is a marvelous product that no one individual fully understands, yet it was accomplished by the organization collectively.

In this first chapter, we are exploring the concept of the black box and how it has contributed to all the great technology we find ourselves surrounded by today. In the remaining chapters, we will dissect some common technology we use in everyday life. We will examine "What's Inside the Black Box".

A perfect example of the black box concept is modern "object based" computer languages. They allow teams of programmers to collectively create insanely complex computer programs. Here's how it works! A programming "object" is a small system of code that performs a function, such as drawing an object on the computer screen. It has certain properties or characteristics that can be set, like size, shape and color. It has certain functions or actions it can perform like draw a circle, make it a little bigger, move it, remove it, etc. One programmer on the team designed this "object" and programmed it. (Or perhaps, one person defined the function, another wrote a detailed specification of how it would work, and a third actually wrote the code.) To do so, he needed to thoroughly understand how it works and how it accomplishes what it does. But he is the only one who needs to understand all that. The other members of the team don't need to know any of the internal details. To use it, all they need to know are the parameters of the object that need to be set, and the actions that can be called for it to perform. They can then use it without any understanding of its inner workings. Collections of programming objects like these are known as software libraries. Any modern computer language has hundreds of software libraries and tens of thousands of pre-programmed objects that programmers can use to accomplish various tasks. Each is a "black box" to everyone except the guy who created it.

If you are young enough to have started dabbling in computer programming after the year 2000, you probably never learned anything but object based programming, and it seems completely natural to you – why would anyone try to design software any other way? But object based programming didn't arrive as the main way to create software until the 1990s! And if you learned to program before that, when programs were viewed as simply data and code, thinking in an object-based way about software seemed strange and very foreign! But it had so many advantages – code can be compartmentalized, encapsulated and reused. Each object of code becomes a black box. Eventually everyone has had to make the transition.

Another great example of black box technology is the modern cell phone. I am going to talk about it here is this chapter, first because it is such a great "black box" example, and second, because I am not going to devote a specific chapter to how cell phones work. They are too complicated. To go into any detail about how modern "smart" cell phones work would fill a whole book by itself.

And yet we all have cell phones. We all know how to use them. And we all have a pretty good idea of what they will and won't do.

But the modern cell phone is an incredible array of technology. The electronics in a cellphone would have filled several rooms just a few years ago. It contains thousands of subsystems, millions of lines of computer code, and the collective work of tens of thousands of people. It is easy to say that no one individual understands more than $1/1000^{th}$ of the collective technology in a cell phone. It contains thousands of black box systems, and is a great example of a black box in total.

To give you some idea of a cell phone's complexity, let's just look at some of the pieces, each of which is a black box even to most of the people who work in the cell phone industry.

There is an audio system, where sound from the microphone goes in, and sound received is sent to the speaker. The sound is digitized coming in from the microphone and it is converted back to analog audio on its way out to the speaker.

There is a radio frequency or RF section that maintains two-way communication with nearby cell-towers. It not only communicates digitized voice in both directions, but also communicates with the cell-tower, identifying itself and associating your phone with your account, and facilitating the handoff between cell-towers as you move about.

In a "smart" web-enabled phone, there are additional RF sections that handle data communication between your phone and the cell tower, and also with your local Wi-Fi network.

There is a computer system, a very sophisticated one, with all of its sub-systems – maintaining your pictures, your apps, your email, Internet access, text messaging, GPS navigation, your contacts, reading material, utilities, maps, etc.

There is a camera system – itself with numerous sub-systems.

There is a system that stores and plays all your music. We could go on. It would take a whole book to go through the basics of how all this stuff works!

But again, to return to our subject of black boxes, no one has to understand how each of the thousands of subsystems in a cell phone actually works. As users, we can treat the whole phone as a black box. As a cell phone designer, you would need to deal with the integration of all these subsystems, but not necessarily know how any individual sub-system works. They'd all be just black boxes. And if you were an expert in cell tower communication protocol, you would know a whole lot about that specifically, and have a fairly good knowledge of RF digital communication. Your area of expertise would not be a black box to you. However, you probably wouldn't know any more than the rest of us do about all the other technology in a cell phone!

Let's look at one more example of the black box concept. I spent most of my own career in the semiconductor industry, where computer chips are designed and manufactured. These devices are built on wafers of silicon using a number of different processes with names like CMOS, NMOS, etc. The entire wafer fabrication process, which takes three to five weeks, is performed in highly specialized facilities referred to as fabs. The fabs and the processes they support are very complex technology. Each fab itself is a little different in capability in terms of the size of individual transistors and characteristics of them.

Each fab publishes a set of "design rules". They also publish computer models for the transistors and other circuit components they can manufacture. The design rules basically define very precisely what the fab can and can't do, and the device models define how the devices built in that fab will function. The design rules and device models essentially say precisely what designs can go into the fab, and what exactly will come out!

Circuit designers are the guys who actually design the integrated circuits. They have no detailed understanding of how the fab actually works. The design rules and device models, however, allow circuit designers to design their circuits without needing to know anything about the inner working of the fab; they just need to follow the design rules established for that fab.

So for the circuit designer, the fab is just a black box. It can be a mysterious place where his designs magically become real. They can be manufactured and mass produced there, but he doesn't need to understand how that happens. As long as he follows the design rules, the fab can remain a black box!

So that's your introduction to the concept of the black box. The rest of this book will explore the inner workings of some common items we use in everyday life. But we probably know them only as black boxes. In each chapter, we'll take one of those black boxes apart and see what's inside and how it works.

Chapter 2 – The Light Bulb

At first you might think a light bulb is pretty simple and not much of a black box. We could even say that a light bulb, since it gives off light, is by definition the opposite of a black box. But it is a pretty good example of what we mean by the black box concept. First - it's something we all use every day and take for granted. Second - while we all know how to use a light bulb, we don't necessarily know how it actually works. And especially today, with so many different kinds of light bulbs, there are a lot of different technologies at work behind the scenes.

So we are going to look at 3 different types of light bulbs: first the incandescent bulb invented in 1879 by Thomas Edison. Then we'll look at the florescent bulb, which has evolved in recent years into the compact florescent lamp or CFL bulb. Finally we will look at the newest type of light bulb – the LED bulb made from light emitting diodes.

We will also want to talk about the efficiency of each of these technologies, as efficiency is what drives light bulb technologies. A large part of our energy use and fossil fuel consumption goes into making electricity, and a large portion of our electricity goes toward making light. The original incandescent light bulb is only 10% efficient, that is, only 10% of the power consumed actually produces light. The other 90% produces waste heat. So improving the efficiency of the light bulb represents a huge opportunity to save energy!

So let's start by looking at the original incandescent light bulb, the one Thomas Edison invented in 1879. When we pass electricity though a long, thin filament, it heats up. In air, it also quickly burns up because of the oxygen in the air. Edison reasoned that by enclosing the filament in a bulb with most of the air removed, that he could heat the filament up to a much higher temperature without it burning up. He experimented for months using different materials for his filaments. His initial success was with a cotton fiber impregnated with carbon particles.

Incandescent light bulbs haven't changed much in the past 135 years. Even today, they still generate light by heating a tungsten filament until it reaches 4,172 degrees Fahrenheit and glows white-hot.

But the incandescent bulb has its drawbacks. As we have already said, it's only a miserable 10% efficient! And its filament, though much improved since Edison's time, still slowly evaporates onto the inside of the bulb, becomes thin and brittle, and eventually breaks after hundreds of times of heating up and cooling down. So its reliability and service life are limited.

For completeness, we should mention one particular type of incandescent bulb here – the halogen lamp. It's like a regular incandescent bulb, but with a small amount of iodine or bromine added inside the bulb. This inhibits the evaporation of the filament, so it increases the life of the filament, and allows the bulb to burn a little hotter and brighter than a conventional incandescent. It's also a small amount more efficient than a conventional incandescent bulb, but not nearly as efficient as the other bulb technologies we are going to talk about.

Today, incandescent light bulbs come in thousands of different sizes, shapes, wattages, socket types, etc. The most popular of these are rapidly being replaced by compact florescent lights, which are both more efficient and more long lasting. So let's look first at florescent lights in general, and then more specifically, at the compact florescent.

People began looking for ways to make light bulbs more efficient almost as soon as Edison invented the original. Florescent lights came into widespread use starting in the 1930s in industrial and office settings, where they could provide bright white lighting for a fraction of the operating cost of incandescent bulbs, but florescent light was generally considered too harsh for home lighting.

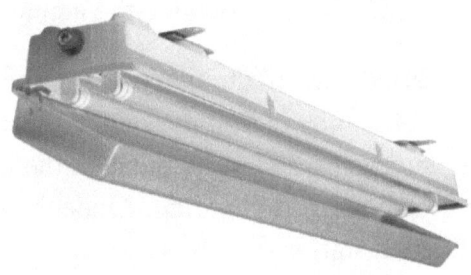

Now let's look at how florescent lights work. The tube is filled with an inert gas, usually argon, at a pressure of about 0.3 atmospheres, and a little bit of mercury which vaporizes when the lamp is operating. The outer walls are covered with a phosphorescent material which is what actually generates visible light.

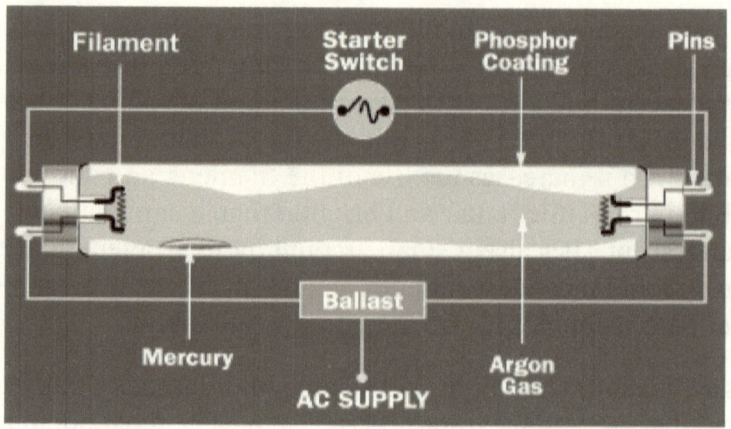

When the bulb is on and operating, electricity flows between the two ends of the tube, exciting electrons in the mercury gas which causes the gas to emit ultraviolet radiation, which is higher in frequency than the visible light that we can see. That's where the phosphorescent coating comes in. It is excited by the ultraviolet radiation, absorbs it, and then re-radiates visible white light which we can see.

There are still a couple of things to discuss about how florescent lights work. One is getting them started, and the other controlling the current after they start. These two things are accomplished by extra circuitry outside the actual florescent tube.

To get the light started, two filaments, one at either end of the tube, heat up when the tube is first turned on. They get hot enough to emit electrons which flow through the tube until the mercury in the tube begins to vaporize. Once the mercury becomes vaporized, a starter switch is opened, the bulb begins to light, and the mercury finishes vaporizing. The heated filaments are only on just long enough to get this process started.

As the bulb warms, the current increases through the lamp. Eventually it would overheat and blow up, if it were not for the ballast. The ballast is a regulator (actually an electrical component known as a chock) which limits the amount of current that passes through the lamp.

All this is obviously a little more complicated than a regular incandescent bulb, but it is over 4 times as efficient as an incandescent bulb at turning electricity into light.

Because of the external ballast, starting circuit, and the straight long tube, conventional florescent lights had serious limitations when it came to using then in our homes to replace incandescent light bulbs.   But around the year 2000, that started to change thanks to automated manufacturing and miniature electronics. Compact florescent bulbs (CFLs) started to become practical. Technology allowed us to wrap the tube around to make the bulb look more like an incandescent. New phosphorescent coatings made the light a softer white. Small electronic starters and ballasts that could be placed in the base allowed them to resemble incandescent bulbs in size and shape.

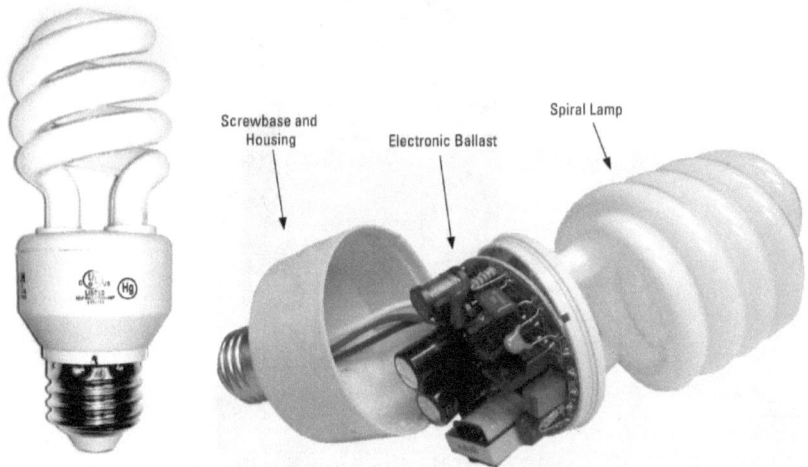

Finally today in 2014, mass production has brought the cost of these lamps down to only perhaps 4 times the cost of an incandescent. With the longer life and much lower operating cost, today CFLs are much more cost efficient than incandescent lamps. And CFLs are being made in more shapes and sizes all the time, so today you can find almost any lamp in a CFL version.

Today, we are also starting to see many types of light bulbs come out in an LED or "light emitting diode" version. Let's see how LED bulbs are made, and then look at their relative cost and efficiency compared with CFLs and incandescent bulbs.

A diode is an electrical component that only conducts electricity in one direction. In 1962, engineers at Texas Instruments discovered a diode technology using gallium arsenide that emitted infrared radiation. Soon the technology was producing visible light as well. By 1968, the semiconductor industry was commercially producing LEDs.

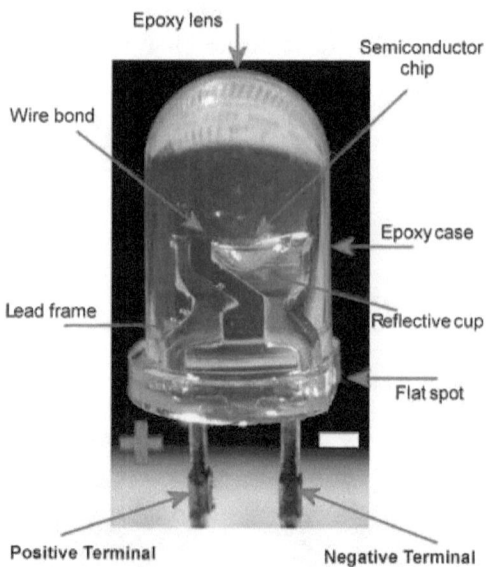

How do LEDs work? All diodes have positive and negative terminals which meet at a junction. An energy barrier called the band-gap is associated with this junction. When a voltage exceeding the band-gap is applied to the diode, current flows through the diode. Electrons take on the band gap energy when they move across the junction, and release that energy when they get to the other side. When that happens a photon is released. If the band-gap energy corresponds to visible light, and if the semiconductor material is transparent, the photon can escape as visible light.

LEDs not only produce light, but they are also very efficient at converting electricity into light, perhaps around 60% efficient compared with 10% for incandescent bulbs and 40% for CFLs.

An individual LED doesn't produce a lot of light. It also doesn't operate from the 110 volt AC we have in our homes. It requires direct current and only 2-3 volts to fully power it. So making a light bulb from LEDs requires a bunch of LEDs and some extra circuitry to power them.

The pictures below show the replacement for a conventional light bulb built with LEDs. There are a number of them required to produce the light of a regular incandescent bulb.

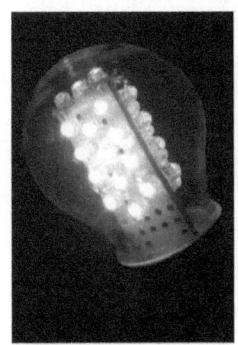 

Looking inside the base, we see the small power supply required to convert the 110 volts AC to the required DC voltage necessary to power the LEDs. And because efficiency is what we are after, that little power supply needs to be very efficient itself or it defeats the whole purpose of using LED technology.

As with CFLs, LED lights are becoming available in all kinds of shapes and sizes. They are up to 50% more efficient than CFLs and last roughly 5 times longer, but their initial cost is still a problem. They are 20 times the cost of an incandescent and 5 times the cost of a CFL. So it takes many years for the initial investment to be paid back in durability and efficiency. But LED bulbs will continue to gain in popularity, especially if the cost continues to decline.

So now you know a little bit about what's inside a light bulb. It's no longer a "black box"!

Chapter 3 – The Modern Flush Toilet

This one is kind of like the light bulb. It's something we totally take for granted, and don't think much about. But it's a very clever device, and it's not very obvious how it works. So we are going to take a look in this chapter.

There are basically two major subsystems to the toilet. The first is the basin itself and the design of the flush mechanism behind it. The second is the tank with a fill system, flush valve, bowl refill, etc. We will look at each of these.

First, let's look at the basin and flush mechanism. It is actually a siphon system.

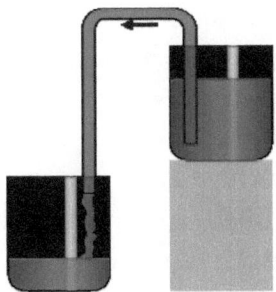

A siphon is a bent tube used to move a liquid over an obstruction to a lower level without pumping. The weight of the liquid in the longer tube pulls the liquid up and though the higher portion.

Siphons are commonly used in irrigations systems, or to drain a liquid out of a container. But they are also at work in every toilet!

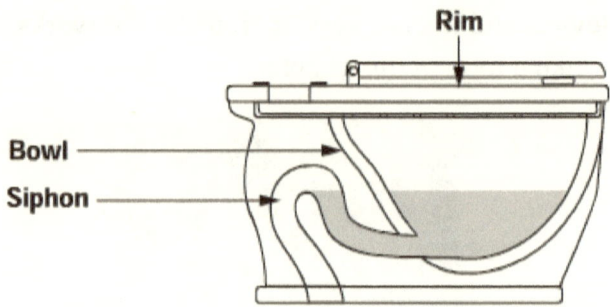

In the toilet's resting position (not flushing), the water in the bowl and siphon tube looks like it does in the picture above. If you slowly add water to the bowl, it would just spill over the edge at the top of the siphon and nothing would appear to change.

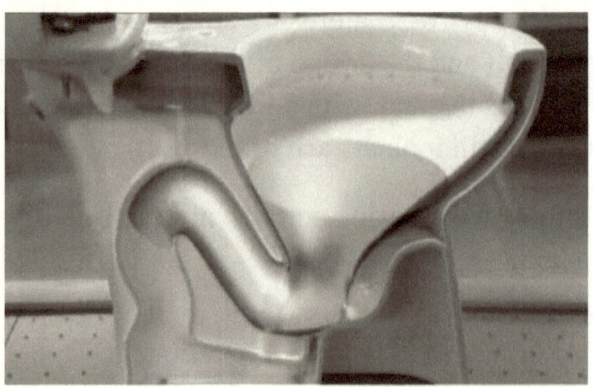

But if you add a lot of water quickly to the bowl, as happens when the tank's flush valve is opened, the top of the siphon becomes completely filled, and siphon action takes over. The siphon action pulls all the water in the bowl and tank out through the drain, emptying the bowl almost completely. The result is a flush of the toilet bowl. This action occurs simply as a result of the physical design of the bowl and siphon.

However, as we said, initiating a flush requires adding a lot of water to the bowl quickly. That's where our second subsystem, the tank and flush valve, come in.

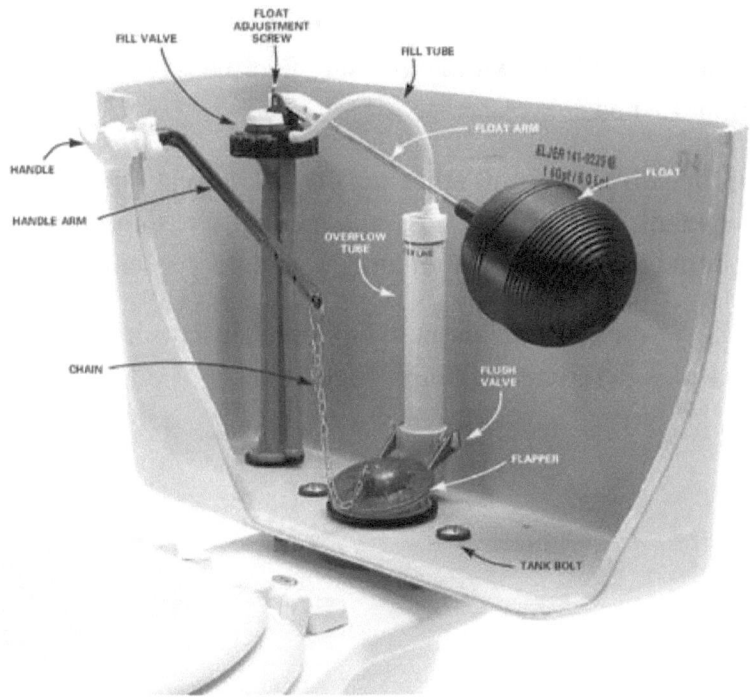

When you flush a toilet, as the photo above shows, all you are doing is pulling up on the chain attached to the flapper at the bottom of the tank. The flapper and the valve seat it rests on are normally closed, holding the water in the tank. Flushing simply means pulling open the flapper, and allowing all the water in the tank to quickly fall into the bowl below. The design of the flapper makes it stay up and open while water is flowing out of the tank, but when all the water is gone, the flapper simply falls back in place allowing the tank to refill.

Now let's look at the rest of the stuff in the tank. The fill valve is at the top of a narrow column attached to the incoming water line. It is operated by a float. When the float is down, as it is when the tank is empty, the valve opens to allow water to fill the tank. As the tank fills and the water and float rise, the float eventually closes the valve when the tank is full. An overflow tube allows excess water to flow into the bowl, if the float doesn't successfully close the fill valve in time. A small fill tube runs from the fill valve to the overflow tube. It adds just enough water to the overflow tube to insure that the bowl refills after being flushed.

All the valve parts and other stuff that can wear out in a toilet are in the tank, and are very easy and inexpensive to replace. So modern toilets are easy to maintain, and very reliable.

So now you know all about the modern flush toilet. If it was a black box before, it isn't any longer!

Chapter 4 – The Microwave Oven

The microwave oven is a black box we all are familiar with, and an appliance we may use multiple times a day. It is a black box in the sense that we know what it does; we know how to set it up, and we know how to get the expected results. But what's inside and how it works is mostly a mystery. So in the chapter, we'll tell you what's inside the black box and how the microwave works.

First we will examine the whole idea of heating food with microwaves. Then we will examine how a microwave oven goes about producing microwaves and applying them to your food.

Microwaves are a form of electromagnetic radiation, like radio waves and light.   In fact, microwaves are higher frequency than radio waves and lower frequency than light, so somewhere in between - actually closer to radio waves. Like all electromagnetic radiation, microwave radiation contains energy and can produce heat, just like the light radiation from the sun warms our planet Earth.

In a conventional oven, heating coils heat the air in the oven, which in turn heats the outside of the food, and then that heat gradually penetrates into the center of the food until the whole thing is cooked.

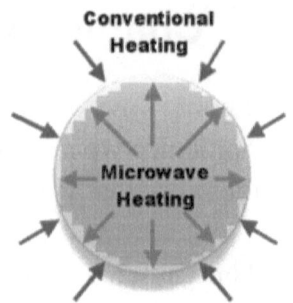

What makes microwave cooking different is that the microwaves are able to penetrate into the food and essentially heat it from the inside. The result, which we all know and love, is that microwaves cook our food much faster!

Why microwaves? We need a particular set of properties when we cook food from the inside. We need the radiation we use to penetrate the food, but we don't want it to interact with and heat up containers like a glass bowl. We need it to bounce off metal, so that we can contain it inside the cooking chamber, i.e. we don't want it leaking out and heating us, for example. It turns out that microwaves are the right frequency of radiation to have all these properties.

So how does a microwave oven produce microwaves to heat our food? Let's look at one from the inside:

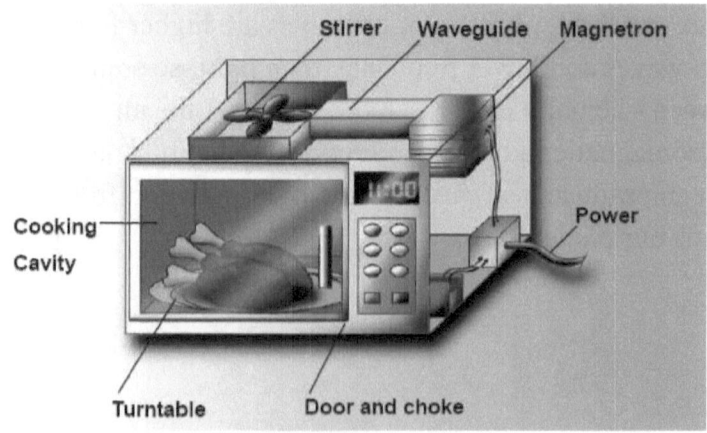

At the heart of a microwave oven is a vacuum tube known as a magnetron. It's what actually produces the microwaves. We will take a look at the magnetron in more detail a bit later. Microwaves, as already stated, bounce off metal, so they can be moved around or steered with a metal tube called a waveguide. The waveguide directs the microwaves to the cooking area. A "stirrer" which is basically a slow-moving metal fan blade is used to scatter the microwaves over the entire cooking area. The metal walls of the cooking area further scatter the microwaves, hopefully producing as uniform heating as possible.

You might be wondering about the glass window through which you can see your food cooking. It has a metal screen behind it which is very effective at preventing microwaves from moving through the glass. They simply bounce off the screen just like they bounce off the metal walls of the cooking area.

Of course, there is a lot more to a microwave oven than what we have talked about so far. There is a control panel where you can set the cooking time and a whole lot of other things. All the timing and other controls are achieved using a microcontroller. We won't talk about that here, but in a later chapter, we will discuss microcontrollers and how smart appliances work.

Let's now talk about the magnetron at the heart of a microwave oven. It's mostly a physical contraption - a vacuum tube in the middle with magnets and metal cooling fins around it. Apply a high enough voltage and microwaves come out of its antenna!

The image below shows the inside of the magnetron tube. You can see why I call it a physical contraption. The shape of the inside of the tube, more than anything else, is what produces the microwaves and determines their frequency.

Below is another picture of the inside of the magnetron. At the center is an electrode that is charged by the power supply to negative 4000 - 5000 volts. Electrons come off that electrode and are attracted to the outer case which is grounded (0 volts). They move in a spiral due to the magnetic field which sets up vibrations or resonance within the chambers inside the magnetron. That resonance is the source of microwave energy.

As electronic devices go, the magnetron is sort of a dinosaur. It's been around for about 80 years, and hasn't changed much over time. It's simple and ultra-reliable. It just puts out microwaves year after year, which is the biggest reason why your microwave is probably the most reliable and long lasting appliance in your home.

By the way, the magnetron is either on or off. It doesn't have any way to vary its power. But for all kinds of reasons, you may want to vary the power setting of your microwave oven. The smart electronics in your oven provides power setting by cycling the magnetron on and off. 50% power is achieved with 10 seconds on, 10 seconds off, 10 seconds on, etc.

So now you know what's inside a microwave oven, and you know at least a little bit about how it works. So one black box isn't a black box anymore!

Chapter 5 – Commercial Electric Power

Let's start with a brief summary of what you probably already know about commercial power. It's produced at power plants, either nuclear, coal powered, natural gas, or hydroelectric. It's also produced to some extent by wind farms and solar panels. It's sometimes transmitted long distances to reach its point of use. It comes into our homes in the US as 60 cycles AC at either 110 volts or 220 volts.

In this chapter, we'll start with a brief introduction to electricity in general, so that we can talk about voltage and amperage. Then we'll look at how commercial power is produced, how it's transmitted and why it's AC or alternating current. And later, in Chapter 15, we'll follow up with how solar energy fits into all this.

Here is some basic information about electricity. Normal physical matter is electrically neutral. That's because the negative charge of the electrons moving around the atomic nuclei and the positive charge within the nuclei balance each other out. But anytime we can "pump" electrons from one place in a material to another, this neutrality is disrupted and we end up with a negatively charged area (with excess electrons) and a positively charged area (with a corresponding shortage of electrons). When this happens, we say an object has an electrostatic charge.

The difference in electrostatic charge between one place and another is called voltage, and is typically measured in Volts. Voltage is similar to pressure in a water pipe. It's the pressure to move electrons within a material until it is once again electrically neutral.

In working with electricity, we use a difference in voltage to move electrons around in wires. The actual flow of electrons is what we call current. Amperes are a measure of current – as in how many electrons per second are moving past a given point. Just like voltage is similar to pressure in a water pipe, current is like the volume of water moving through the pipe per second.

It takes energy to pump water up-hill, and in similar fashion, it takes energy to pump electrons from one voltage to another. At a dam on a river, we can get energy out of water stored above the dam by running it through a turbine. The amount of energy we get is a combination of the pressure times the flow of water. So Power = Pressure x Flow Volume.

Here too the analogy between water and electricity is valid. A battery is a source of energy. It uses that energy to create a voltage difference and maintain it while current flows. The energy produced per second by a battery is the voltage times the current - what we call Wattage. So Power (watts) = Voltage (volts) x Current (amps), just like our example with water above.

We still need one more term to finish our discussion: Resistance - the opposition to the passage of an electric current through an electrical component. In our water analogy, resistance would be like the friction between the water and the pipe, or in a more extreme case, a significant restriction in the pipe. Resistance is measured in Ohms, which is the ratio of voltage to current in an electrical component.   So an important concept is:

Resistance (ohms) = Voltage (volts) / Current (amps)

This relationship between resistance, voltage, and current is known as Ohm's Law. If you know any two of the three, you can calculate the third. For example, how much current is being drawn by this 3 ohm light bulb?

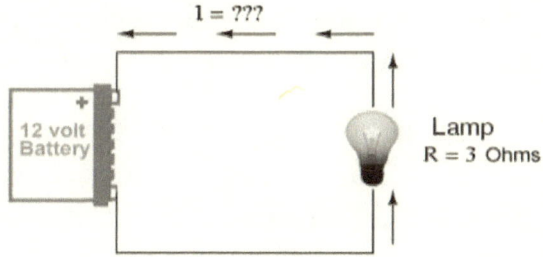

In Ohm's law math, current is usually referred to as "I", as it was called "intensity" before it was called current. So the answer is I = V/R or 12 volts/3 ohms = 4 amps, which we found using Ohm's Law. Let's go one step further.   What is the power or wattage of this light bulb? Power (watts) = voltage x current = 12 volts x 4 amps = 48 watts.

Now let's change subjects slightly and talk about commercial electricity. The power from a battery is direct current or DC. The electrons flow from the negative terminal to the positive terminal. But as you know, commercial electricity is alternating current or AC. The current flow reverses direction 60 times per second. So let's see why we use alternating current and then discuss the implications.

For commercial power, we frequently have to transport electricity over long distances – hundreds of miles between the generators and consumers. Some power is lost in the process of transmission due to the resistance of the wires.   The power lost in the wires is Power = V x I, where the V in this case is the voltage drop over the length of the wires. But from Ohm's Law, V = I x R, so the power loss is (I x R) x I which is I^2 x R. So the power loss is proportional to the square of the current! The implication of this is that to transmit power over a long distance, with a minimum amount of power loss during transmission, we need the current as low as possible. We need to transmit power (which again is V x I), so for any given amount of power, the lower we make the current, the higher we need the voltage to be. That's why power is transmitted at very high voltage. Typical transmission line voltage is 230,000 volts!

But how do we get the voltage up and the current down. That's what transformers are for. So next we'll look at how transformers work, and then see what the implications are for commercial power.

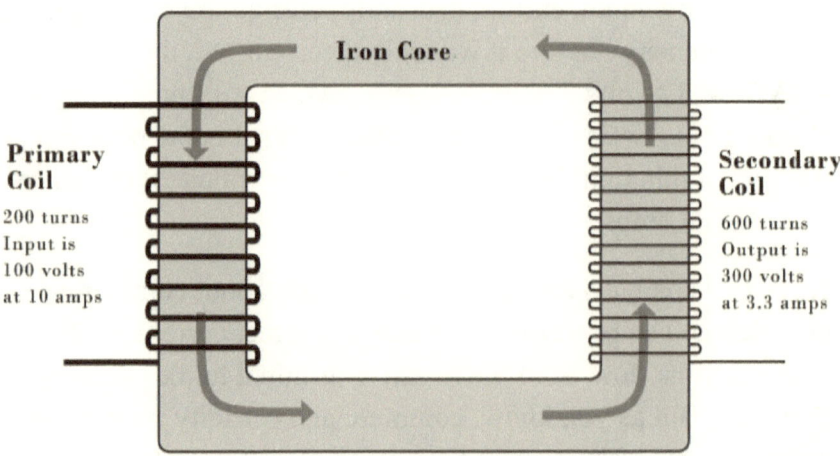

A transformer has two coils or windings, both wrapped around an iron core. The ratio of turns between these two windings determines how voltage and current change between the primary (input winding) and the secondary (output winding). So in our example above, the secondary winding has three times as many turns as the primary winding. If the input voltage on the primary is 100 volts, the output voltage on the secondary will be 300 volts. Transformers are very efficient, so roughly the same power going into the primary comes out of the secondary. But power = voltage x current. So if the voltage at the secondary has increased three times, then the current at the output must be 1/3.

Transformers do exactly what power companies need. One can raise the voltage and lower the current for transmission, and then, at the power's destination, one can do the reverse – lower the voltage and raise the current. But transformers work because of the magnetic coupling in the iron core of the transformer, and that coupling only works with alternating current. So that is the reason why commercial power is AC. Direct current like we get from a battery cannot go through transformers, and therefore, cannot be efficiently delivered by power companies.

Transformers come in all different shapes and sizes. Below, on the left, is a big transformer like what might be on the receiving end of a 230,000 volt transmission line.   On the right is a neighborhood transformer that might serve four homes. Power to this particular transformer comes from underground cables.

The drawing at the top of the next page illustrates the process in total. The power plant generates alternating current. It's transformed by transformers to very high voltage for transmission to cities.

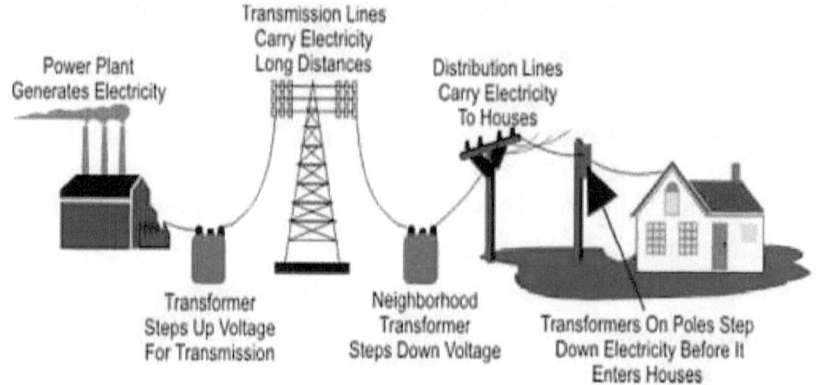

Transmission Lines
Carry Electricity
Long Distances

Power Plant
Generates Electricity

Distribution Lines
Carry Electricity
To Houses

Transformer
Steps Up Voltage
For Transmission

Neighborhood
Transformer
Steps Down Voltage

Transformers On Poles Step
Down Electricity Before It
Enters Houses

At a local substation, like the one shown below, the voltage is reduced for local distribution within a city, and then the voltage is reduced again by neighborhood transformers before it enters our homes.

Now let's look at how electricity is produced at a power plant. We'll use a coal-fired power plant as our illustration, but whether it is coal, oil, natural gas, or nuclear powered, these heat based power plants work the same. First, heat is used to boil water, producing steam under pressure. The steam is used to turn the blades of a turbine, which in turn drive a generator which produces alternating current. The generator looks and acts very much like an electric motor. It has a combination of magnets and coils, but instead of using electricity to turn a motor, the turning rotor in a generator produces electricity.

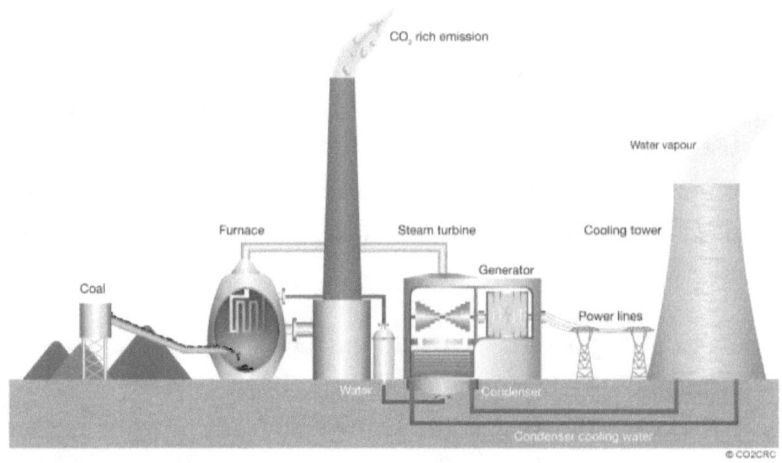

The speed of the generator must be carefully regulated so it produces 60 cycles AC and its output must be synchronized with the rest of the power grid it is hooked up to.

The steam must be turned back into water which requires large cooling towers that are characteristic of power plants. And for all the gas and coal based plants, there is a lot of CO_2 (carbon dioxide) gas emitted into the atmosphere.

Hydroelectric power plants are very similar; however, the turbines are driven by falling water instead of heat and steam.

To complete our discussion of commercial power, we need to look at what goes on inside your home. Most of the stuff in your home, such as lights and small appliances, don't use a lot of power. They are powered by 110 volts. The four things that do use a lot of power (central air conditioner, water heater, clothes dryer, and electric stove) all operate on 220 volts. Why? Because at 110 volts, they would draw twice as much current to get the power they require. Even at 220 volts, the current drawn by these power hungry appliances require extra heavy wiring and special connectors.

So now you might ask: "why do we have two different voltages in our homes?" It's a good question. In Europe, everything runs off 220 volts. But in the United States, 110 volts is used for lights and small appliances because it is safer – a much lower shock hazard than 220 volts.

Many electric appliances in your home can use AC directly. The 220 volt ones all use AC, as do items like light bulbs, small heaters, irons, and hair dryers. But most electronic devices, like your computer or television require DC (direct current) for power, usually at several different voltages. So "power supplies", which are circuits built into these devices, convert the AC into the DC voltages required to operate the electronics.

So now you know about commercial electric power. It's no longer a black box.

Chapter 6 – A Car's Automatic Transmission

It's a black box even to most car mechanics, except for those specializing in transmission repair. An automatic transmission is a wonderful piece of technology that has been refined and fine-tuned for the last 70 years. Most of us drive a car that has one, but we basically take it for granted, and certainly know nothing about how it works. In this chapter we will de-mystify the automatic transmission.

Let's start by talking about what any transmission does, whether manual or automatic. The name "transmission" comes from its role in the transmission of power from the engine to the wheels. But it needs to do more than just that.

It's the characteristics of the gasoline engine that make transmissions necessary to begin with. Gasoline engines, once started, need to keep turning. They operate within a range of speeds – a few hundred revolutions per minute (RPM) at idle to maybe 4000 RPM. They are most efficient in a narrow range – perhaps 1500 – 2500 RPM. But we want our car to operate smoothly and efficiently from 0 miles per hour (MPH) to perhaps 80 MPH. The conflicts between what the engine wants and what we need presents some problems. Transmissions, whether manual or automatic, solve these problems using a system of gears.

In contrast to the gasoline engine, let's look at the electric motor, which in 2014 is starting to replace gasoline engines. Electric motors can be powered from a dead stop. In fact, an electric motor draws maximum current and provides maximum torque at a dead stop. (Torque comes up several times in this chapter. It is simply the rotational force being transmitted from the engine to the wheels.) As an electric motor speeds up, the amount of current it draws and the torques it generates gradually declines.

So it can accelerate quickly at low speeds, and at the same time be energy efficient at high speeds. These are the characteristics we want in the engine that powers our car. The electric motor has these characteristics inherently and is just starting in 2014 to replace the gasoline engine as a car's power source. Electric motors can directly drive the wheels of our electric car – no transmission is required.

But almost all cars today are still powered by gasoline engines, and they require a transmission to solve their shortcomings and allow them to behave more like electric motors.

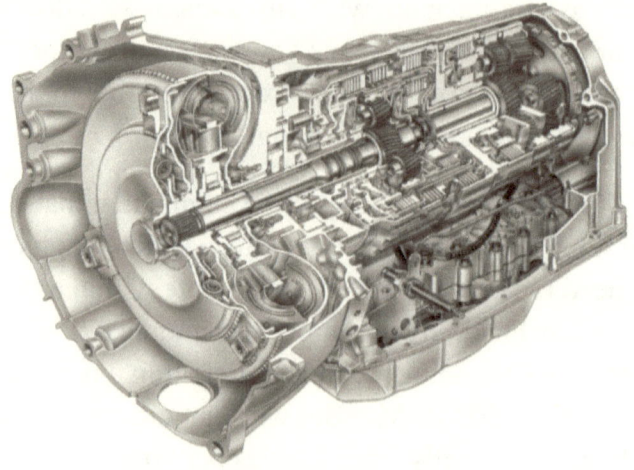

A cutaway view of an automatic transmission.

So what does a transmission (manual or automatic) do? First, in conjunction with a clutch (or in the case of an automatic transmission, a torque converter), it allows the engine to run at idle while the car sits at a stop. As the engine speeds up and the car starts to move, it allows the engine to run at a near optimum speed, while the wheels of the car move at slow speed. Using gears to accomplish this, it also increases the available torque and acceleration at low speeds. As the car continues to accelerate, gears must be changed to keep the engine running near its optimum speed. And of course, we also need the transmission and its gears to give us reverse, because gasoline engines (unlike electric motors) only turn in one direction.

So the subject of this chapter is the automatic transmission. How does this magic piece of technology accomplish all this? And of course, what's inside the black box?

To understand the inner workings of an automatic transmission, we need to examine its various subsystems. First, at the heart of an automatic transmission is its planetary gear set. Let's look at that first.

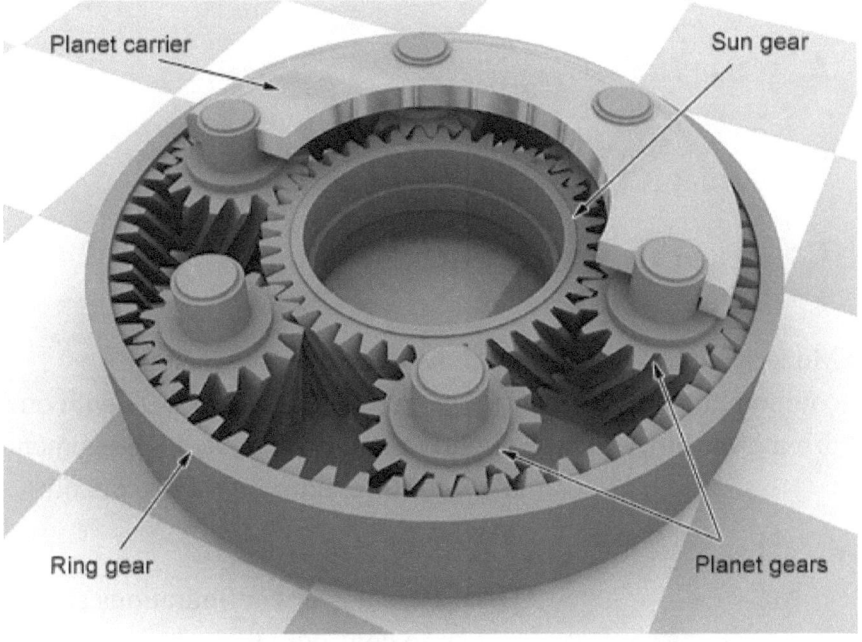

A planetary gear set, like the one above, consists of three parts – an inside gear called the sun gear, an outside gear called the ring gear, and a set of planetary gears in the middle that drive the third piece – the planet carrier.

The cool thing about a planetary gear set is that the gears are always engaged in the same configuration. Yet depending on which gear is being driven, which gear is providing the output, and which gear is held stationary, a number of different net gear ratios can be achieved, including reverse. (A gear ratio, by the way, is the ratio of the number of turns of the input to one turn of the output.)

The table below shows some different configurations of a single planetary gear set which are capable of producing low gear (A), high gear (B) and reverse gear (C). Note that the gear ratio for reverse is negative.

	Input	Output	Stationary	Gear Ratio
A	Sun (S)	Planet Carrier (C)	Ring (R)	3.4:1
B	Planet Carrier (C)	Ring (R)	Sun (S)	0.71:1
C	Sun (S)	Ring (R)	Planet Carrier (C)	-2.4:1

Modern automatic transmissions frequently have either a compound planetary gear set or two planetary gear sets with one driving the other. These slightly more complicated arrangements

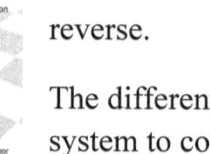

allow for 6 or more forward gears plus reverse.

The different gear configurations and the system to configure them is our next subject.

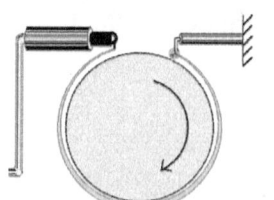

At the left you see a planetary gear with a band around the outer ring gear. Tightening the band locks the ring gear and prevents it from turning. The band is actuated by a piston which pulls the band tight around the ring gear, locking it in place. These pistons and bands in an automatic transmission are how the different gear configurations are achieved.

Next we need to look at what drives the pistons which in turn actuate the bands and change the gears within an automatic transmission.

The maze of hydraulic circuits you see in the picture below is the control system for the pistons and bands. It is basically an analog computer running completely on hydraulic fluid. Various sensors like engine RPM, throttle setting, drive shaft RPM, engine temperature, etc. all send messages to this system using hydraulic valves and pressure, and this maze of hydraulic lines uses that information to decide when to activate the various bands to select the appropriate gear. While this whole hydraulic computer idea sounds very strange, it works extremely well!

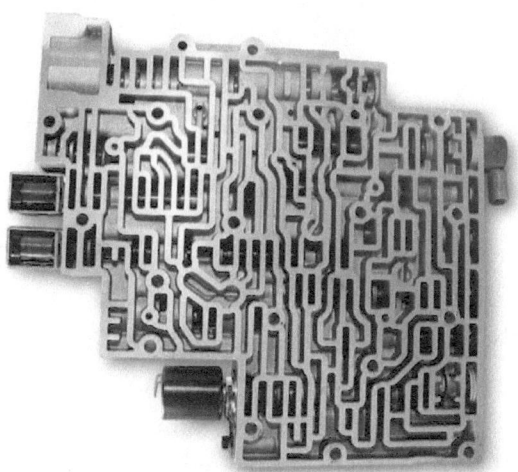

As you might expect, modern digital electronics is gradually moving in to replace this strange analog computer, but it is happening surprisingly slowly. That's probably because this old hydraulic system has been refined and perfected over the past 60 years, and today is amazingly reliable. "If it's not broke, don't fix it!" comes to mind, but electronics are gradually taking over parts of this system, or in a few cases, completely replacing it.

The next part of the automatic transmission we need to discuss is the torque converter. We might have chosen to start with it, as it is the thing that is attached directly to the output of the engine. Why is it needed and what does it do?

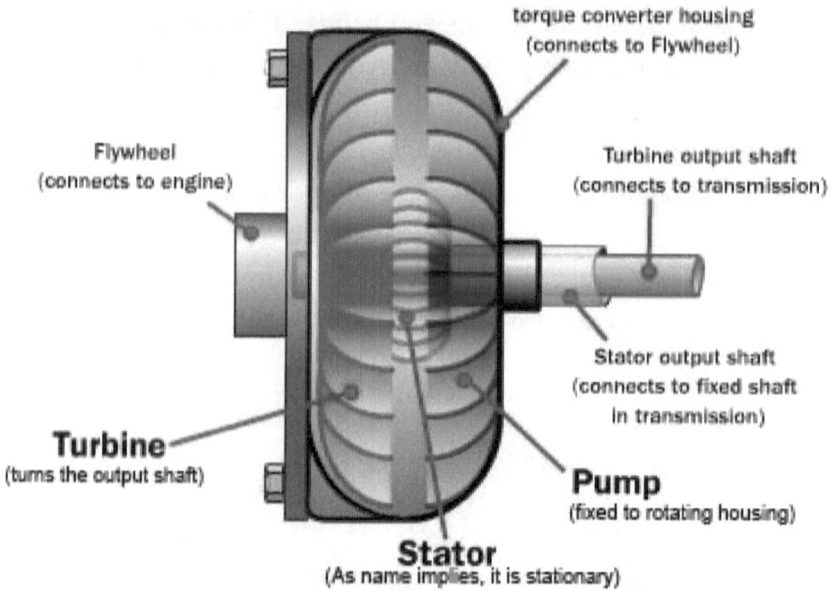

A torque converter is a type of fluid coupling, filled with transmission fluid, allowing the engine to spin somewhat independently of the transmission. It takes the place of a clutch in a manual transmission. If the engine is turning slowly, such as when the car is at a stop with the engine idling, the amount of torque passed through the torque converter is very small, so keeping the car at a stop requires only a foot on the brake pedal.

But when the engine speed is increased to above an idle, a large amount of torque is transmitted through the converter to the rest of the transmission.

What makes the torque convertor different than a regular fluid coupling is a thing in between the pump and the turbine called the stator. It is fastened to the housing of the transmission, so it does not turn as the pump and turbine do. Instead it redirects fluid returning to the pump from the turbine in such a way as to multiply the torque transmitted to the turbine whenever the speed of the turbine is less than that of the pump. Since maximizing torque at lower speeds is one of the objectives of the transmission as a whole, this feature is a nice compliment to the gears.

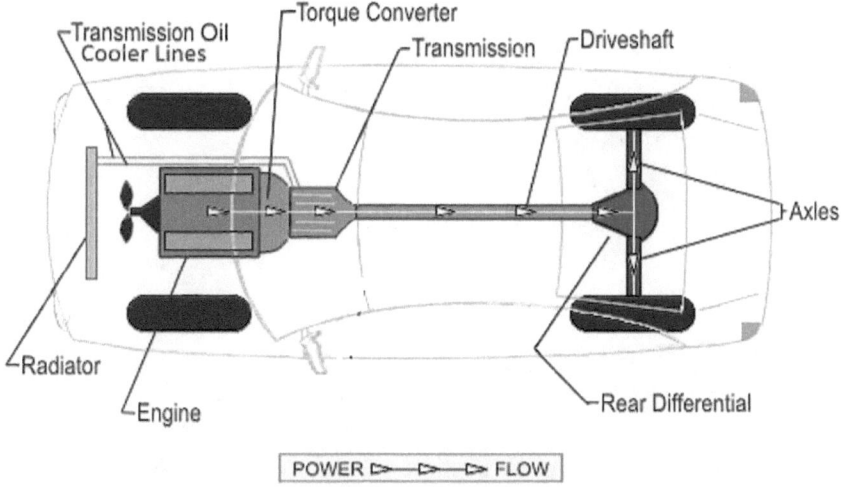

Now let's put it all together. Power moves from the engine to the torque converter, from the torque converter to the planetary gears in the transmission, which are selectively configured depending on speed, throttle (gas petal), and other factors to select the proper gear. Then power moves from the back of the transmission through the drive shaft to the differential and from there to the rear wheels.

So now you know at least a little about how an automatic transmission works, and what's inside the black box.

Chapter 7 – GPS Navigation

GPS navigation tools are suddenly everywhere. Our smartphones have them built in, our cars have them, or if they don't, Garmin, TomTom and many other companies build add-on systems to help us plan our routes, and be guided effortlessly to our destination. But what is GPS and how does it work?

Let's start by defining GPS. It stands for Global Positioning System, and it's a satellite-based navigation system made up of a network of 24 satellites placed into orbit 11,000 miles above the Earth by the U.S. Department of Defense. GPS was originally intended for military applications, but in the 1980s, the government made the system available for civilian use. GPS works in any weather conditions, anywhere in the world, 24 hours a day. There are no subscription fees or setup charges to use GPS.

GPS satellites circle the earth twice a day in very precise orbits and transmit information to earth. GPS receivers take this information and use triangulation to calculate the user's exact location. Triangulation is essentially a process of narrowing down one's possible locations until a single position is determined. The GPS receiver compares the time a signal was transmitted from a satellite with the time it was received. The time difference tells the GPS receiver how exactly how far away the satellite is, because the signals are travelling at a known speed, namely the speed of light. Now, with distance measurements from four different satellites, the receiver can determine its position, not only latitude and longitude, but also elevation. And the whole thing is accurate to within a few feet.

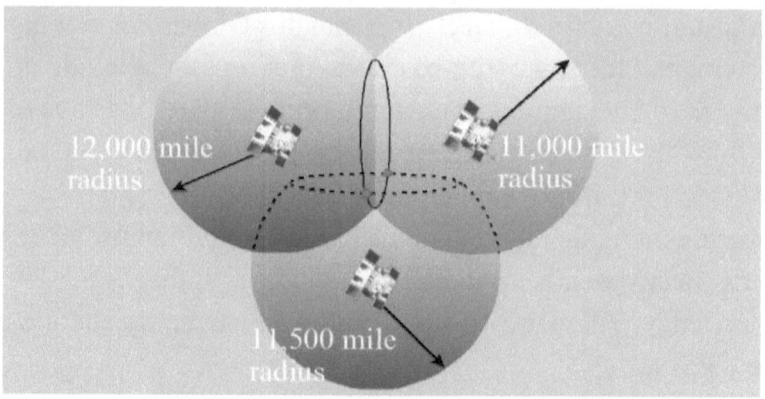

The image above shows how knowing the distance from just three satellites has narrowed the position of the receiver to just two points. Adding a fourth narrows down the position to a single point in space and improves the accuracy. The receiver can now calculate latitude, longitude, and altitude anywhere in the vicinity of Earth.

Let's now look in a little more detail about how the system works to determine a user's position. Then we will look at how a GPS navigation system uses that information to help you navigate.

Conceptually, measuring your distance from 4 different known positions to determine your own position is straight forward. But since the 24 satellites are themselves flying around the Earth at about 6000 miles an hour, it's a little more complicated. First, timing is extremely important. So each satellite has a super-accurate atomic clock on board. A ground control system makes sure that the satellites know exactly where they are and that their clocks are accurate.

The user's GPS receiver also has a clock that is synchronized with the satellite clocks. So everything in the system knows exactly what time it is, and every satellite knows its own exact position at any one moment.

So each satellite sends out its exact position and a clock timing signal every few fractions of a second. The GPS receiver receives these signals from at least four satellites. It can calculate its distance from each satellite by measuring the time delay between when the clock signal was sent and when it was received.

Needless to say, there is a lot of computer power required to make this whole system practical. Even a little handheld receiver has to be able to measure the transmission time delays down to fractions of a microsecond to calculate the distance to the satellite, as the incoming signals are traveling at the speed of light. And they have to be able the process all this data to calculate the longitude, latitude, and altitude from that data.

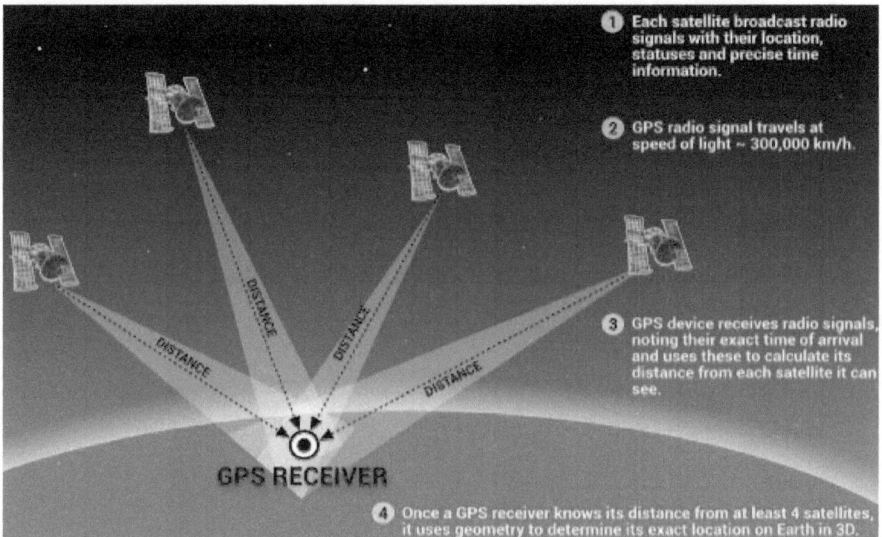

1. Each satellite broadcast radio signals with their location, statuses and precise time information.

2. GPS radio signal travels at speed of light ~ 300,000 km/h.

3. GPS device receives radio signals, noting their exact time of arrival and uses these to calculate its distance from each satellite it can see.

DISTANCE
DISTANCE
DISTANCE
DISTANCE

GPS RECEIVER

4. Once a GPS receiver knows its distance from at least 4 satellites, it uses geometry to determine its exact location on Earth in 3D.

We know now how the GPS system can find out exactly where we are at any moment, but how do these little systems help us navigate to a new location? Let's now look at how a navigation system works.

If you have used one of these systems, you already know what they do. You tell them where you want to go, they have built-in maps, and look at different routes. They pick the best route, and then guide you step by step through that route, using both a display screen map, and also voice instructions. We will now look in a little more detail about how all this is accomplished.

A lot of data is required to accomplish all this. A car's navigation system uses four different databases to decide the route and present it to the driver:

- Background data - rivers, shores, mountains, railroads, contours, etc.
- Road network - road configurations, connections, conditions, attributes, such as speed limits, etc.
- Site information data - building names, addresses, representative structures, etc.
- Voice data - voice guidance

These data bases are provided and updated regularly by the system's manufacturer. They are stored in memory chips built into your navigation device.

The system then uses these various data sources to guide you to your destination, by going through the following steps:

1. Find your current location on the map in the road network database, and identify nearby roads, connecting routes between roads, as well as attributes such as one-way streets and intersections.

2. Find and set the driver's selected destination. This is accomplished by applying information about the destination in the site information database to find the destination in the road network database, and again for the destination, identify nearby roads, connecting routes between roads, as well as attributes such as one-way streets and intersections.

3. After confirming the current location and the location of the destination, the system searches for routes that connect the two locations. Road connection status, intersections, left/right turn conditions, etc. are determined. This process actually involves three different maps, each with different levels of detail. There is the fundamental map, which is used to navigate through a city. There is a wide area map, which shows major routes between cities. Then there is a detailed map, used only near the starting point and destination.

4. The system now must determine the recommended route. To do so, it must use its background data and road network databases to give each possible route a score. Positive contributions to the score come from being most direct, shortest, fewest road changes, best speed limits, etc. Negative contributions to the score come from narrow roads, one-way streets, frequent congestion, mountainous terrain, etc. The route with the highest score is recommended.

5. After the route is decided, driving directions commence. Information on buildings and roads along the route is confirmed and displayed on the screen. Voice guidance is simultaneously provided, using information obtained from the voice database.

6. Additional information is displayed on the screen. While route directions are being provided, information on the locale through which the car is driving is displayed. In this way, roads, buildings, topographical features, etc., along the way are identified. In some cases, the car's navigation system can

confirm which roads are congested on the basis of real-time information received from various organizations concerning accidents, road construction, etc. The system can also provide constant updates on your speed, estimated arrival time, etc.

GPS and the navigations systems that use it are truly amazing technology which we more or less take for granted today. They rely on an incredible amount of computer power which would have been completely impossible only a few years ago.

So now you know how GPS and your car's navigation system work. They are no longer a black box.

Chapter 8 – Smart Appliances

Fifty years ago, appliances used to have an on/off switch and maybe a temperature control or something similar. But today, almost every appliance has a huge array of features, settings, automated functions, programmable options, etc. Today, we have "smart appliances"!

All these new features and options are part of the computer revolution. Microprocessors, small computers on a single chip, were first introduced in 1974. Today they come in a wide range of capabilities and price. The ones that are fastest and possess the most features power our personal computers, tablets, and smartphones.

The smaller, slower, and less powerful ones are called microcontrollers. They cost as little as a dollar and they are built into practically every appliance and small electronics item on the market today. But don't think, because they are smaller and slower than the ones in your personal computer, that they don't do much. Today, even the simplest microcontrollers can do as much as the best computer in the world could do 50 years ago!

In this chapter, we will explore microcontrollers and how they work to make today's appliances "smart".

Microcontrollers have inputs and outputs. The inputs read data from the outside world, like the position of a switch, or the temperature. The outputs can turn things on or off, and display information. Microcontrollers can combine various inputs and process them to make decisions about the outputs. They do this by following a predefined program or list of instructions. And they can process several million instructions per second, which frequently gives them seemingly amazing capabilities.

The individual instructions given to the microcontroller by the program stored inside it are pretty simple, like:

1. Configure pin 7 as an output.
2. Configure pin 2 as an input.
3. Read pin 2 - this gives us a 0 or 1.
4. If it's a 0, wait 2 seconds and make pin 7 = 1.
5. If it's a 1, wait 5 seconds and make pin7 = 0.

The microcontroller simply steps though its list of instructions one by one, completing each is less than a microsecond. When it finishes the list, it goes back to the beginning and starts over again. But with these types of simple instructions, we can gradually add a lot of functionality and intelligent decision making into the microcontroller, and from there, into our appliance.

You might logically ask how the code above actually does anything. We saw the microcontroller look at pin 2, see it was a 0, and decide to wait 2 seconds and make pin 7 = 1. Well pin 7 might be hooked to a transistor. Making it 1 turns that transistor on, this in turn closes a relay (switch) which turns on a motor or heating element. In this way, the microcontroller is able to control how an appliance actually behaves and what it does.

The next step in understanding how microcontrollers can make an appliance "smart" is to look at a specific example. Since we already talked about the microwave oven in chapter 4, and promised to talk about smart appliances, we are going the use the microwave oven to illustrate how microcontrollers make smart appliances.

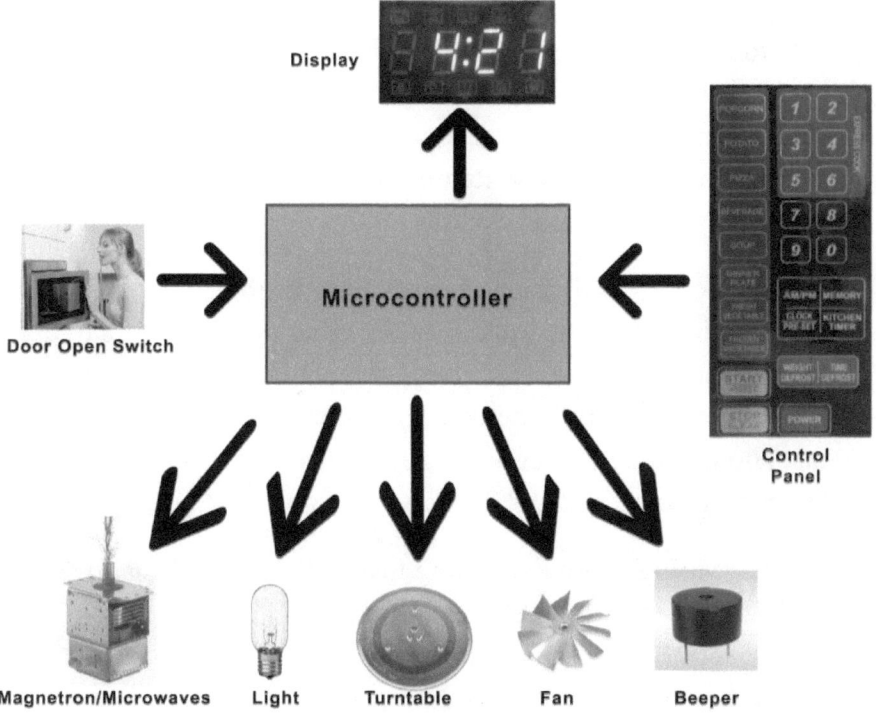

Display

Microcontroller

Door Open Switch

Control Panel

Magnetron/Microwaves   Light   Turntable   Fan   Beeper

The picture above shows how a microcontroller manages the various activities within the microwave oven. The arrows show the flow of information. So, the microcontroller gets an input from a switch telling it if the door is open or closed. It outputs information to the display to show the time. It receives inputs from all the buttons on the front control panel. And with those inputs, it makes decisions about all the other stuff, like when to turn on the magnetron, turn the turntable, and send out a beep when the food is done.

One thing that we haven't talked about is timing. Microcontrollers are very good with time management. They maintain a very accurate internal clock, with counters that keep track of hours, minutes, and seconds. That's why it's easy for them to display the time. They typically have several timers build right into the chip to allow them to manage several tasks at once, and keep track of the timing for each of them.

You might wonder why a microcontroller would even need to be able to execute millions of instructions per second. Part of the reason is that they really can only do one thing at a time, yet we want them to sometimes manage 50 tasks at the same time. We achieve this by rapidly switching back and forth between tasks many times per second.

So the microcontroller is kept very busy keeping the clock's time, keeping the cook time up to date, keeping the display up to date, turning the magnetron on and off, watching the control panel for a new input, watching the door switch etc. It has to do all of this fast enough and often enough that we perceive everything happening concurrently.

But everything is not happening instantaneously. What if the door is opened while the magnetron is on? Will the microcontroller see it fast enough to turn off the magnetron before microwaves come spilling out into the room? We don't even want to find out. So in this case, the switch is wired directly to disable the magnetron without even waiting for the microcontroller to detect that the switch is open.

So the microcontroller isn't infinitely fast, even if it can perform millions of instructions per second. But it does enable the appliance to do many things it couldn't do otherwise. For example, your microwave probably has a button to pop popcorn. How does your microwave know how the pop popcorn? There are hundreds of thousands of instructions stored in the permanent memory of the microcontroller. One of them is a recipe to pop popcorn. Another is a recipe for defrosting. Another is a recipe for reheating. All these recipes are available to you at the push of a button, which makes your microwave seem pretty smart.

Today smart appliances are everywhere. Sewing machines are another perfect example of the "smart appliance" which gains new functionality from an embedded microcontroller. Its microcontroller is pre-programmed with hundreds of different stitches. It can automatically sew button holes; it can thread itself, and all sorts of other things.

So now you know more about your microwave oven and a tiny bit about computerized sewing machines. And you also know some things about microcontrollers, and how they add features and programmability into many appliances. "Smart appliances" are no longer black boxes!

Chapter 9 – Electric Motors

Most people don't realize how many electric motors they have in their homes and cars. Let's take a look – I put my own count in parenthesis:

- bathroom fans (3)
- ceiling fans (8)
- air conditioner - inside fan motor, outside fan motor, compressor (6)
- shaver (2)
- electric tooth brush (1)
- computer - ventilation fan, CPU cooling fan, power supply fan (6)
- printer - paper feed motor, head scanning motor (6)
- refrigerator - outside fan, inside fan, compressor, ice maker dispenser, ice tray ejector (5)
- microwave - turntable, microwave stirrer (2)
- dishwasher - pump (1)
- garbage disposal (1)
- washer - pump, tub (2)
- dryer (1)
- hand mixer (1)
- blender (1)
- most clocks with hands (4)
- most watches (5)
- car - windshield wipers, ventilation, power seats, fuel pump, power windows, power mirrors (12)
- portable fans and heaters (2)
- vacuum cleaner (2)
- sewing machine (1)
- electric drill (2)
- electric saw (2)
- electric yard tools - mower, blower, trimmer (2)
- toys (3)
- pool pump (0)

I counted 81 electric motors in my house and cars! Chances are I missed a half dozen in my count!

Electric motors are everywhere, and we tend to take them for granted. But there are many different kinds of electric motors, and most of us don't really know how they work. So this chapter is devoted to demystifying the electric motor.

Let's take a very quick look at the history of the electric motors. Scientists first began experimenting with magnetic coils and made simple motors in the 1820s. Electric motors were first introduced commercially in the 1870s, but with no commercial power yet, they operated from batteries and were very limited in what they could do.

In the 1880s, however, things started to change very rapidly. As commercial power became available, electric motors suddenly found all kinds of important applications, like powering elevators and trolley cars. And they revolutionized agriculture and manufacturing over the next 20 years!

As we already said, there are many different kinds of electric motors. Here is a list:

Types of Electric Motor

DC Motors	AC Motors	Other Motors
Shunt motor	Induction	Stepper motor
Separately Excited	Synchronous	Hysteresis motor
Brushless DC		Reluctance motor
Series Motor		Universal motor
Permanent Magnet DC		
Compounded		

We aren't going to talk about every kind of motor in this chapter, but we will describe three kinds: the shunt DC motor is where we will start, as it is a good introduction to electric motors in general. Then we will talk about both AC motors – induction and synchronous, as both are found all over your home.

Before we try to look at specific types of electric motors, let's define some things that most electric motors have in common. First there is a **rotor** – also known as the armature, it's the moving part which turns inside the motor. Then there's the **stator** – the stationary part. There are **coils** on both the rotor and stator, which generate magnetic fields when energized with current. The stator coil is sometimes called the field winding. Then there is the **commutator** which switches and sends power to the rotor as it turns. Its contacts are called **brushes**, though they are seldom actually made of brush material.

We'll see how this all works by looking at the shunt DC motor. A shunt DC motor runs on direct current, like what you get from a battery, and it has its rotor and stator coils (also known as windings) hooked up in parallel, as shown is this drawing:

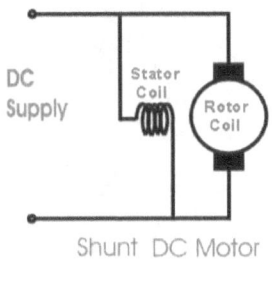

Shunt DC Motor

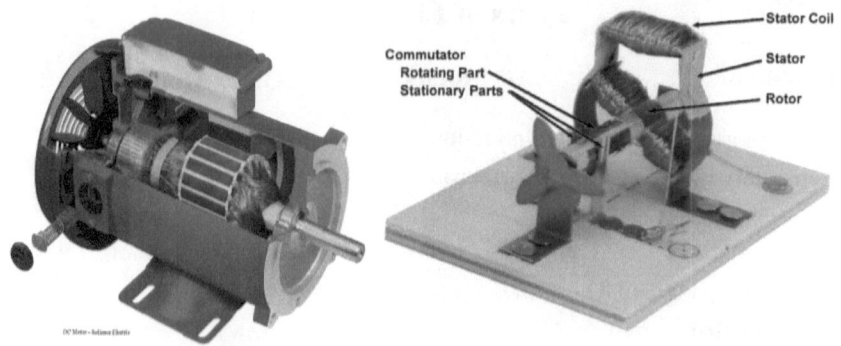

If we try to look at a commercial DC shunt motor, like the one on the left, it's hard to see how it works. So we are going to look at simple version, an educational kit, shown on the right.

The wire coil at the top is the stator coil which produces a magnetic field in the metal piece that goes through it and wraps around the rotor. The rotor sits in the middle and has its coil wrapped around both sides of the center. They produce another magnetic field which moves about with the rotor. The two coils are wired in parallel as was shown in the previous drawing.

The little triangle at the front end of the rotor is just there so that you can see the rotor turn. Just behind it though is the commutator. It has to two stationary contacts (brushes) and two rotating contacts and transmits power to the moving rotor.

Now to make our motor turn clockwise as we are looking at it in the picture, let's connect a battery to the coils such that both magnetic fields on the left would be north poles and would push away from each other. As the rotor moves clockwise, the north pole of the rotor would be attracted to the south pole of the stator on the right side. But because of the commutator, just as the rotor reaches the middle of the right-side stator, the rotating part of the commutator switches the polarity of the current in the rotor coil. Suddenly the rotor is again repelled by the stator and attracted to the stator on the other side. As this process continues, the rotor revolves completely around, repeating the process over and over.

Here is another image to help show how the rotor is turned by the magnetic fields.

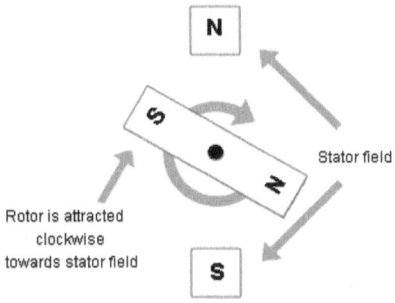

All motors work in similar fashion, with magnetic fields attracting and repelling to drive the rotor around, but they achieve this result in a number of different ways.

Our shunt DC motor used direct current, but most of the motors in your home (and all large motors) run on commercial AC power. So let's look at the AC induction motor. That's the motor in your washer and dryer, your pool pump, and your air conditioner.

The AC induction electric motor was invented by Galileo Ferraris and the famous scientist Nichola Tesla in 1887. Below is a picture of their original motor.

The nice thing about the induction motor is that it is very efficient and reliable. And it doesn't need a commutator, as no current flows directly into the rotor. (That's good, because commutator parts are the first thing to wear out in an electric motor.) Instead, current is induced into short loop coils in the rotor by induction. What that means is that the magnetic field in the stator coil, which is constantly changing due to the alternating current, is able to induce a current in the coils of the rotor. This induced current in the rotor coils produces its own magnetic field and the two magnetic fields work together to drive the motor.

For the induction motor to work, the stator coil (also known as the field winding) must produce a rotating magnetic field when AC current is applied. The drawing below shows how three different coils on the stator produce a rotating 4 pole magnetic field. Induction motors may have 4 poles, but 6, 8 or even 10 poles are sometimes used in induction motor designs. The primary coil produces an alternating magnetic field. The smaller secondary coils cause the magnetic fields inside them to lag behind the main field. The result is a rotating field!

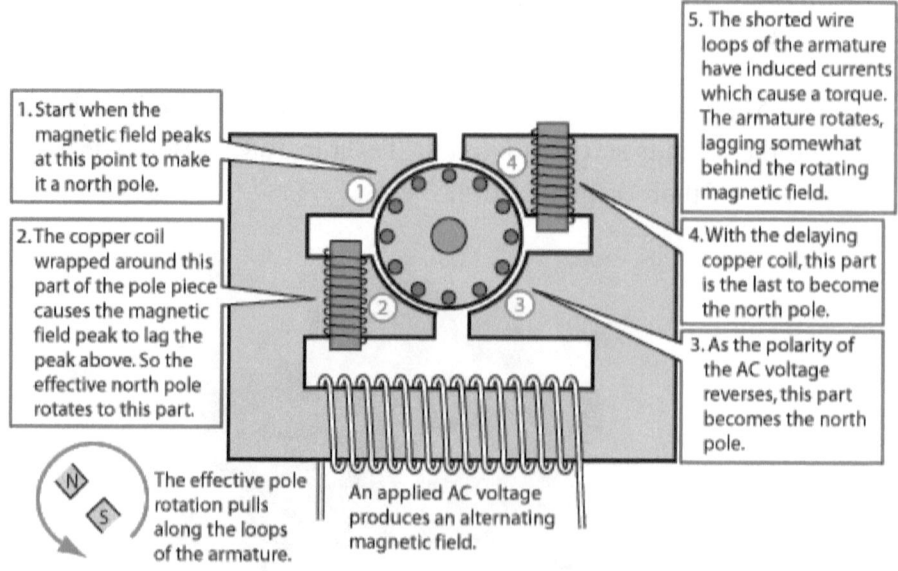

5. The shorted wire loops of the armature have induced currents which cause a torque. The armature rotates, lagging somewhat behind the rotating magnetic field.

1. Start when the magnetic field peaks at this point to make it a north pole.

2. The copper coil wrapped around this part of the pole piece causes the magnetic field peak to lag the peak above. So the effective north pole rotates to this part.

4. With the delaying copper coil, this part is the last to become the north pole.

3. As the polarity of the AC voltage reverses, this part becomes the north pole.

The effective pole rotation pulls along the loops of the armature.

An applied AC voltage produces an alternating magnetic field.

The induced current and resulting magnetic field in the rotor basically causes the rotor to attempt to rotate with the rotating magnetic field in the field winding. However, it never actually does rotate at the speed of the rotating magnetic field. It's always slower. The difference in speed is what causes the induction to work in the first place. And that speed difference increases as the work load of the motor increases. So its speed is variable and asynchronous, meaning not synchronized with the alternating current.

The synchronous AC motor is the other common AC motor. It is perfectly synchronized with the alternating current. They are usually small motors used where precise constant speed is required, like electric clocks, timers and other devices where correct time is required. Below is a typical synchronous timer motor. Motors like these typically have a reduction gear set built into them, so that even if the rotor itself is turning 60 times per second, the output shaft might only be turning once per minute.

In a synchronous motor, the stator or field winding produces a rotating magnetic field similar to field winding in the induction motor. The rotor in the synchronous motor, however, is different than the induction motor. It doesn't have any coils. Instead it has either permanent magnets or a slug of steel that gets an induced magnetism. Either way, there is no slippage as there was with the induction motor. The rotor stays exactly synchronized with the rotating magnetic field generated by the field winding.

So these synchronous motors turn at a precise fixed speed, tied to the 60 cycle AC current in our power lines. The power company precisely controls the AC so that it has exactly 60 cycles each second. This is in part to help make synchronous motors accurate as clocks and timers.

So now you know a little bid about electric motors, and how the most common types of electric motors work. They are no longer black boxes!

Chapter 10 – Heat Pumps
including Air Conditioning and Refrigerators

Your home/car air conditioning system and your refrigerator are both heat pumps and have a lot in common. In this chapter, we will explore how heat pumps work, and then look at the specifics of how they work in your refrigerator and air conditioner.

All heat pumps exploit the fact that a large amount of heat is consumed when something turns from a liquid into a gas; then conversely, a large amount of heat is released when the gas turns back into a liquid.

Although water isn't used in heat pumps, we'll illustrate this principle using water, since we are all pretty familiar with water. As you know, water boils at 212 ° F. It takes 100 calories to heat one gram of water from 32 ° F (freezing) to 212 ° F (boiling). But it then takes an additional 540 calories for that gram of water to go from liquid at 212 ° F to steam at 212 ° F! So a relatively huge amount of heat is consumed in just making the transition between liquid and gas. This heat is called the latent heat of evaporation. And all substances have a latent heat of evaporation. Water has a particularly high one.

While water is used to cool some things such as a nuclear power plant, its boiling point is too high to be useful for cooling our home or our food. We need something that boils at a very low temperature and gets cold when it uses up a bunch of heat to evaporate. We use the term Freon to refer to the general class of compounds that are useful as refrigerants. It's actually a brand name for a whole range of compounds.

Freon refrigerants seem to have a way of leaking into the atmosphere over time, and they are all harmful to the atmosphere, in that they break down ozone and increase the amount of ultraviolet radiation that reaches the ground. For that reason, we have been changing the refrigerants we use over time and getting rid of the worst offenders. So I'm not going to talk about any specific refrigerant. These changes to our refrigerants are also why Freon is a lot more expensive than it once was!

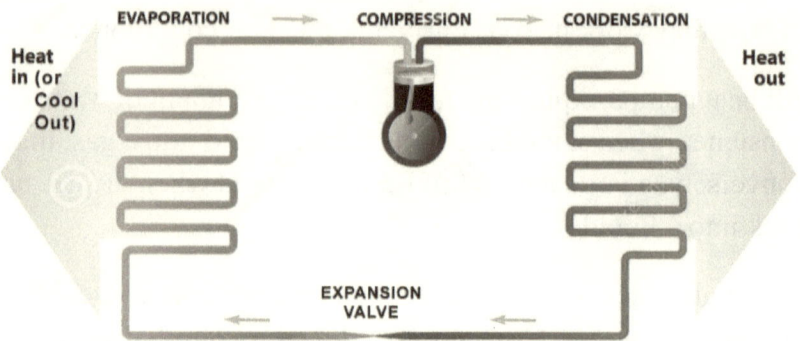

So let's now look at the basic principles of heat pumps. They are closed loop systems in which a compressor pumps a gaseous refrigerant under high pressure until it liquefies. When it liquefies, it gives up its latent heat of evaporation and gets hot. The hot liquid is pumped through a heat exchanger, which in conjunction with a fan transfers that heat into the air. The liquid cools in the process. It then flows still under pressure to an expansion value where it is allowed to expand into the lower pressure evaporator. There it turns back into a gas, and in the process consumes a lot of heat, which makes it get cold. It then goes through a second heat exchanger and fan where it cools air. It is actually absorbing heat from the warmer air, but as a result, it is cooling the air. Finally, the now warmer refrigerant returns to the compressor, where it is again compressed back into a liquid. This process continues and in effect pumps heat from the evaporation side to the condensation side.

Now let's take a look what this heat pump looks like in your refrigerator and talk about how it creates a 0 ° F freezer in one compartment and a roughly 37 ° F refrigerator in the other compartment.

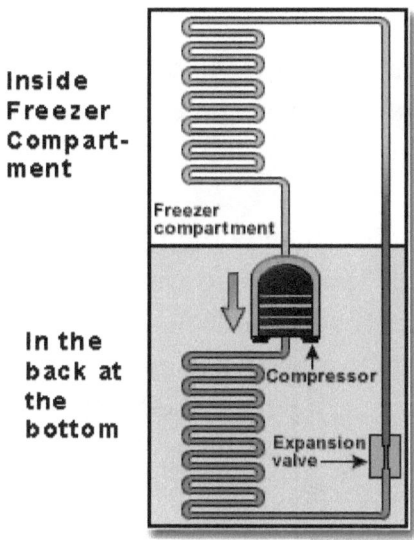

Inside Freezer Compartment

In the back at the bottom

Freezer compartment

Compressor

Expansion valve →

This is what the same diagram looks like in your refrigerator. The evaporation heat exchanger is in the back of the freezer compartment, and the compressor, expansion valve, and condensation heat exchanger are in the back at the bottom.

In the photo on the left you can see the heat exchanger inside the freezer. It's at the back and usually covered up with a plastic panel. In the photo on the right, you are looking at the lower portion of the back of the refrigerator with the cover panel removed. The big back case is a sealed unit that contains the compressor. The fan on the far left is blowing air across the condensation heat exchanger to remove the heat being pumped out of the freezer.

So that's how the freezer gets cold. A thermostat controls the freezer temperature by cycling the compressor on and off to regulate the temperature.

What about the refrigerator portion? There is also a fan that blows air from the freezer compartment into the warmer refrigerator compartment and then back into the freezer compartment. It too is regulated by a thermostat and runs just enough to keep the refrigerator at its desired temperature.

Now let's look at where the heat pump parts are in a home air conditioner.

Some units have everything contained in one box that sits on the roof, but we will look at the two part system with an outside unit that sits on the ground and an inside air handler unit that sits in a closet or in the attic.

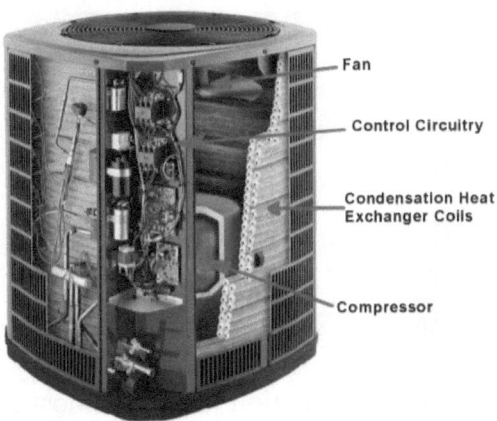

The outside unit contains the compressor and the condensation heat exchanger coil and fan. Two Freon lines run between the two units – one with compressed liquid, and a return line with gas returning to the compressor.

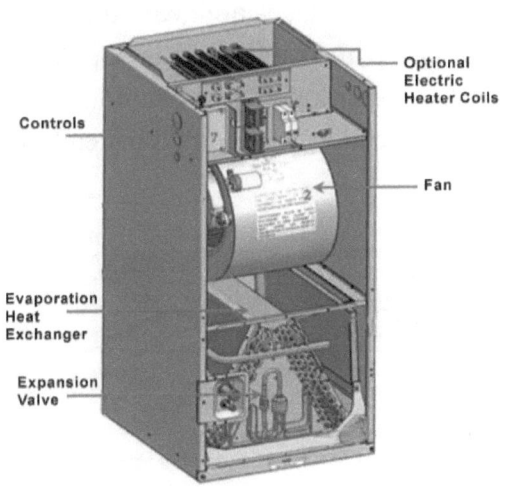

The inside air handler contains the expansion valve, evaporation heat exchanger, and the fan that blows cool air throughout the house.

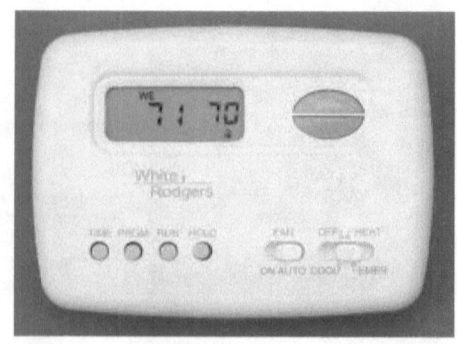

One last item we should mention is the thermostat. It's kind of the brains of the whole thing. It controls both outside and inside units and turns the compressor and fans on and off to maintain the set temperature. Most thermostats today contain a microcontroller where temperature can be programmed to change with the time of day, etc.

I have been using the term heat pump in its general sense, which is to say that all of these devices are heat pumps. But one particular type of home air conditioner is popularly referred to as a "heat pump". Air conditioners that can both cool and heat your home are specifically called heat pumps instead of just air conditioners.

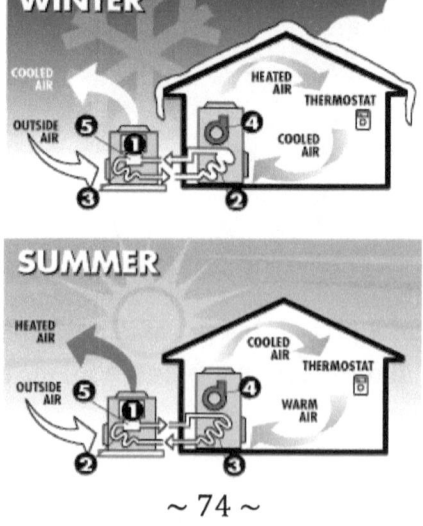

We have been looking at the home air conditioner so far in the cooling mode, but with "heat pump systems", they can work in reverse as well, where the outside coil cools, and the inside coil heats. This is accomplished using a set of reversing valves which cause the whole system to run in reverse. This reversing valve is controlled by the thermostat, where switching to "Heat" instead of "Cool" causes the reversing value to engage when the compressor is turned on.

Generally, heating and cooling heat pumps are used only in very warm climates with limited need for heat. That's because they are designed to be most efficient at cooling, and are far less efficient as heaters. And for this reason, they frequently need to have extra electric heating coils built into the inside air handler to help with the heating.

So now you know about heat pumps, air conditioning and refrigerators. They are now longer black boxes!

Chapter 11 – Television

Television is a huge subject. The technologies involved are everywhere around us – in the TVs themselves, but also in computer monitors, camcorders, and even our cellphones can record video. It's so pervasive, you might be tempted to think it's fairly simple, but there is a mind-boggling array of technology developed over the past 100 years that makes modern television what it is.

We are going to examine this subject over the next two chapters. In the first one we'll examine how the original television came about and how it worked, both in terms of how the signal reached us and how the TV displays it. Then in the second chapter, we will look at how it has evolved in recent years – why it's gone digital, what high definition is, how the sound has improved, and now we even have 3D television.

I first saw television as a child in about 1950. Only one house in our neighborhood had a television, and everyone got invited over to see this great new technology in action. As shown in the picture above, the screen was small, the picture was black and white, and it was grainy and a little snowy. Still, it was amazing, as the only place we'd ever seen moving images up until then was at the movies.

Now let's see how this original version of TV worked. We'll start with how the picture was transmitted. Then we will talk about how your television converted that transmitted signal into a picture.

Today we are used to thinking about digital pixels, e.g. your computer monitor might be 1024 pixels across by 768 pixels high. However, early television was analog, not digital. In the U.S., it had 525 lines vertically, but horizontally, there were no distinct lines, and thus, no pixels. The picture was built line by line from an intensity signal, where minimum intensity represented black and maximum intensity represented white.

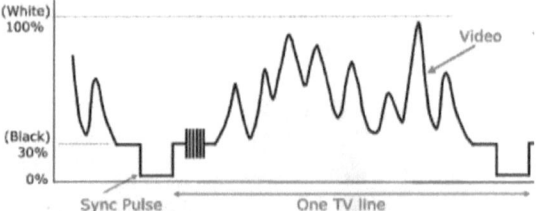

A synchronization pulse was sent to indicate that one line was complete and that information for the next line was starting. It was a synch pulse in the sense that it kept the picture synchronized with the signal being received.

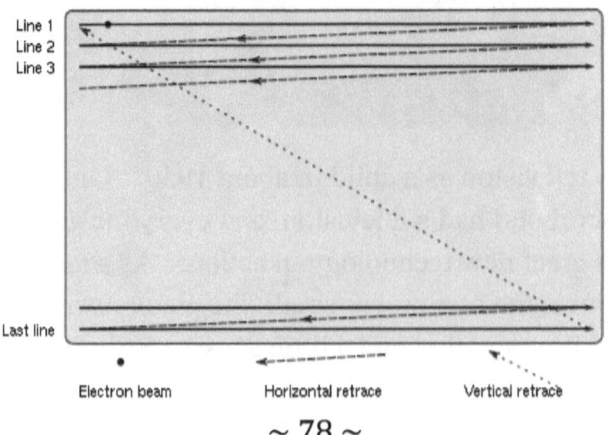

This process continued for all 525 lines, and then the process was repeated, so that the whole picture was sent 30 times per second. Actually, there is one more detail we didn't mention yet. The picture was "interlaced". That means the odd lines were all sent first, and then all the even lines were sent. This interlacing process significantly reduced flicker in the resulting picture.

Now let's look at what your old television set did with this incoming signal to turn it into a picture.

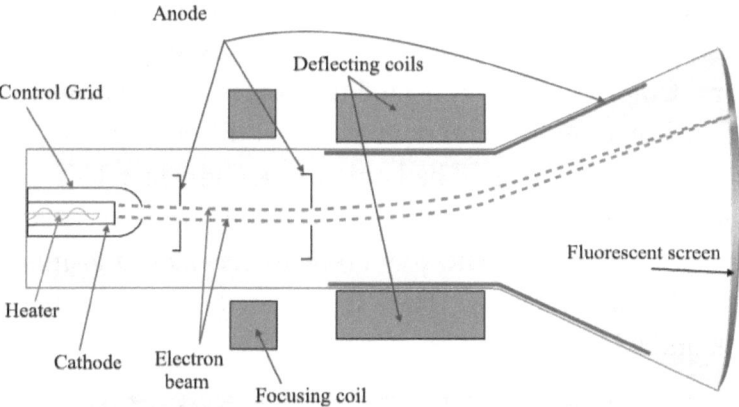

While we won't go into all the details of how the signal was received and processed, we will discuss how it actually became a picture on the face of the television set. A large vacuum tube, called a cathode ray tube, sent electrons out in a tight beam toward the front face of the television. There they produced light when they struck a florescent screen. The intensity signal that we mentioned on the previous page as used to control the intensity of the electron beam and therefore the brightness of the florescent screen. But to actually form a picture, the beam needed to be steered.

Steering was accomplished with deflection coils. As one line was sent, the horizontal deflection coil would sweep the beam horizontally across the front of the screen. With the completion of a line, the synchronization pulse told the vertical deflection coil to step down one line. Horizontally, the beam needed to be brought back to the starting point. During this process, known as horizontal retrace, the electron beam needed to be temporarily shut off or blocked, as the retrace was not part of the picture. This line by line horizontal sweep and step down for each line all happened 525 times for each picture and the whole thing was repeated 30 times per second.

The process we have just described was how the original black and white television worked from roughly 1945 until 1965. Of course, what everybody was waiting for was the introduction of color. Color really complicated the process, first because it involved sending the picture in at least 3 different colors, but also because most people still had only black and white TVs which couldn't interpret the color picture. During the early 1960s, there was a great deal of debate about how to best start transmitting color television pictures. They still needed to be viewable on a black and white TV.

A transmission system had to be worked out whereby the picture intensity signal was still there for black and white TV receivers, but then there was extra information transmitted just for the color receivers that told the set how that intensity was distributed over the three colors. And of course, since many TV programs were still only broadcast in black and white, a color TV had to be able to display black and white shows as well. It really complicated the whole process.

When a standard for broadcasting color was finally agreed upon, networks began transmitting some shows in color. They didn't really amount to much until 1965. The quality of color broadcasts varied widely. I remember, if you were lucky enough to have a color set, you showed it off by inviting your friends over to watch Bonanza on NBC.

Bonanza clearly had the best color picture on TV, and everything else was judged relative to Bonanza.

Now we need to discuss how the color was actually displayed.

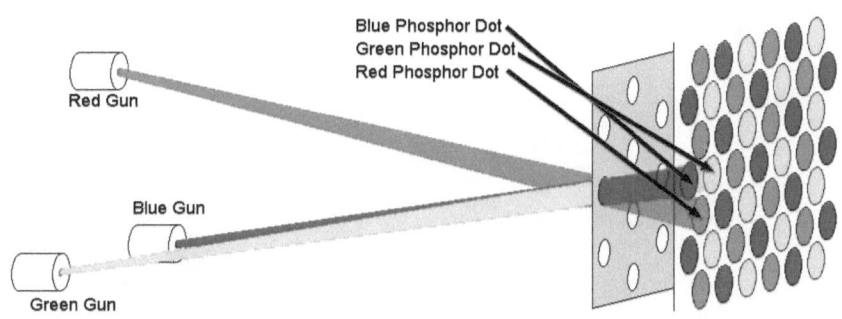

Picture is made up of approx. 300,000 phosphor dots of each color!

A few things needed to be added to the inside of the picture tube to make color possible. First, we needed three different electron guns to send separate beams of electrons for red, green, and blue colors. Next, we needed red, green, and blue phosphor dots painted onto the inside surface of our picture tube. Finally, a metal plate with holes in it, called a shadow mask, was placed just behind the phosphor coated screen. It allowed the red beam to hit the red phosphor, the green beam to hit the green phosphor, and the blue beam to hit the blue phosphor, but prevented them from hitting one another.

This was the state of television from about 1965 until 1985. Picture tubes got bigger, squarer, and better. Broadcast quality improved significantly, and color TV became the norm.

Then in the 80's and 90's, new technologies started to emerge. Television went digital with a big improvement in picture quality. High definition TV came into existence and more recently became the norm. These new developments in television are the subject of our next chapter.

Chapter 12 – Television – Part II

This chapter is about all the wonderful changes that have occurred in television since about 1985.

Color cathode ray picture tubes never got bigger than 40" diagonal, but in the '80s, rear projection TVs got much bigger. Sound systems got much better and most programs were broadcast in stereo. Also, the electronics continued to get better so that the pictures were more stable, and the colors stayed aligned (or converged) with each other.

But just as with the conversion to color, the U.S. was late in starting to broadcast high definition (HDTV) and converting the television signal to digital. The reason was the same as it had been for color conversion: compatibility with existing broadcasting standards. This time the problems were ever bigger than converting to color, as neither digital broadcasting nor high definition could be made compatible with the old analog standard definition broadcasts.

In the U.S., the initial conversion to digital was done in conjunction with the introduction of HDTV, but that was not a requirement. Some other countries that started broadcasting HDTV earlier did so using analog. But in the US, digital HDTV sets became available in 1998, and satellite and cable broadcasters were both starting to supply digital HDTV content by 2002. The HDTV channels were broadcast completely separate from the standard definition analog TV channels; again, there is no compatibility between the two broadcast standards.

The completion of conversion of television to digital was completed in 2009, when all standard definition television broadcasts in the US were required to go digital. So now everything is digital today, but high definition is still broadcast on its own channels.

Today we have digital HDTV instead of analog standard definition. Let's take a second to see what the differences are.

Digital TV results in a huge improvement in picture quality all by itself. Analog is plagued by all kinds of distortion and interference which all disappear when TV is broadcast digitally. For example, snow, grainy pictures, and ghost images all go away with digital - the picture is either there or it's not. Color distortion also disappears. Compared with analog, the digital picture is always perfect.

High definition is another huge improvement in picture quality. Standard definition TV, as we have said earlier, has 525 lines. The analog signal has a resolution about equal to 700 pixels across. Some of this resolution actually overlaps the screen a little, so standard definition TV is generally regarded to be equivalent to 640 pixels wide by 480 pixels high. In comparison, HDTV is 1920 pixels wide and 1080 pixels high. So HDTV has 6.75 more pixels and detail than standard definition!

So now we have seen all the great changes in how television is broadcast. Let's next look at how the television display itself has changed for the better.

We mentioned rear projection TV at the beginning of this chapter. In the late 80s, Texas Instruments invented digital light processing or DLP. This is a process by which a million or more microscopic mirrors on a single chip of silicon can project a high definition image onto a screen. These DLP sets represented a huge improvement in projection TV, starting around year 2000, but big LCD and plasma screens began appearing soon after that, and have pretty much taken over the television market today. These new flat screen technologies are bigger, brighter, lighter, and cheaper to produce that anything that came before them!

We won't look at plasma, but let's look at an LCD TV and see how it works.

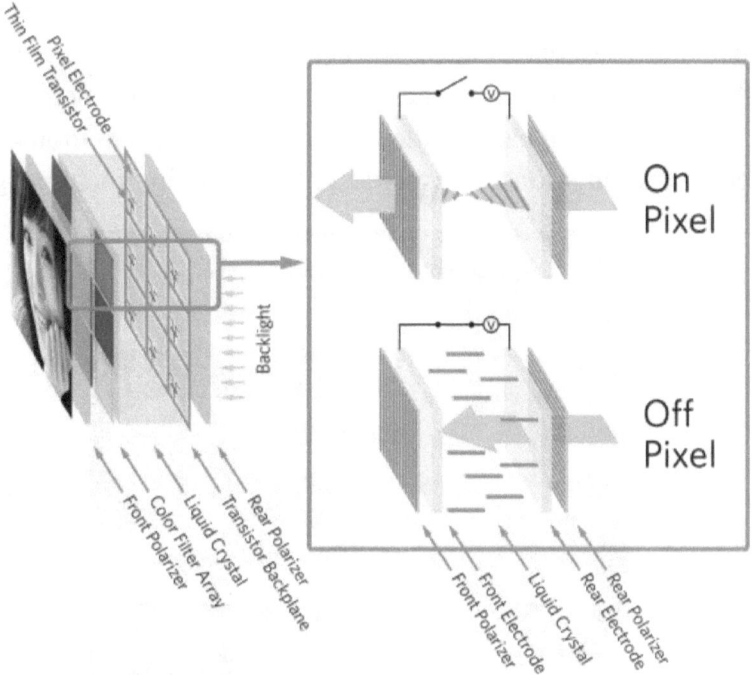

LCD is short for liquid crystal display. LCD panels are typically composed of two sheets of polarized material with a liquid crystal solution between them. Light must be polarized correctly to go through a polarized material, which in this case means being vertically or horizontally polarized. When no electric current is passing through the liquid crystal, the crystal rotates the polarity of the light by 90° to match the second polarized sheet, and light passes through the entire "sandwich. But when an electric current passes through the liquid, it causes the crystals to align so that light isn't rotated and can't pass through. Think of each crystal as a shutter or gate, either allowing light to pass through or blocking it out.

After passing through the front-most polarized pane, the light then passes through a color filter that leaves it red, green, or blue. Each cluster of red, green and blue makes up one pixel on the screen. By selectively illuminating the colors within each pixel, a wide range of hues can be produced on the large display.

This is a sample of the pixels on an LCD screen. There are typically 1920 pixels (three colors each) across the width of the screen and 1080 pixels down the height of the screen on a 1080P high definition set. The digital video processor in the set is capable of lighting each of these millions of pixels in thousands of color combinations in response to the incoming picture signal.

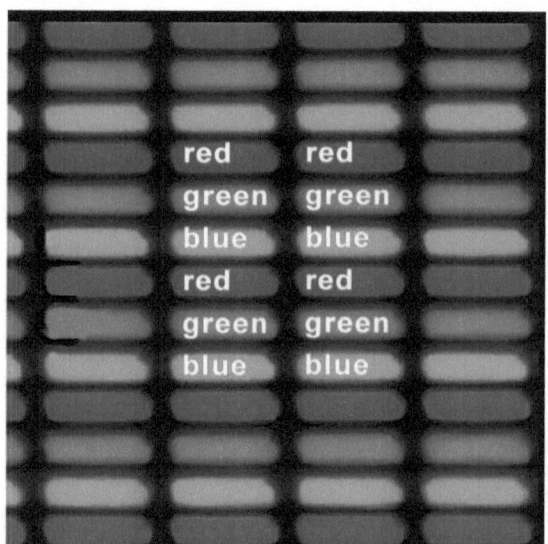

LCD TV has excellent contrast ratios, meaning it can produce a very bright picture, yet the screen is very black where you want it black. LCD TVs today are backlit with large number of LEDs (light emitting diodes), which provide extremely bright, uniform screen lighting – another significant improvement in picture quality.

The last subject we'll cover is 3D TV. Whether you're watching 3D at the movies or on a 3D television set, the concept for 3D is always the same. One picture is shown to only your right eye, and another is shown to only your left eye. They switch back and forth fast enough that it appears you are seeing both images all of the time.

Currently 3D TV is accomplished by polarizing light coming from the screen in one direction for the left eye, and then rotating the polarization 90 degrees for the right eye. Infrared signals transmitted to the required glasses send the right eye image to the right eye, while the left eye image is sent to the left.

Time will tell whether 3D TV will really take over. Many people find the glasses a nuisance. If not, we'll have to wait for holographic TV to get real 3D. But don't hold your breath! Holographic TV probably won't be here any time soon.

As we said at the beginning of Chapter 11, TV is a huge subject. By necessity, we've skipped over a lot of details about how TV works, and we haven't talked about some of the related technologies like video recorders and camcorders. But at least you now know a little about the history of television technology, and a little bit about how it works. Your television set is no longer a black box.

Chapter 13 – Websites and the Internet

Here is another amazing technology we all take for granted today. But how does the Internet actually work? In this chapter, we'll first look at the big picture – what is the Internet, how do the major pieces work, and then, specifically, how do websites work. Then we'll look at how websites are created.

Let's start with the really big picture first. The Internet is an incredibly large network of interconnected computers. Hundreds of millions of these computers are users viewing web pages or reading emails, while millions of others are servers supplying the webpages or distributing emails.

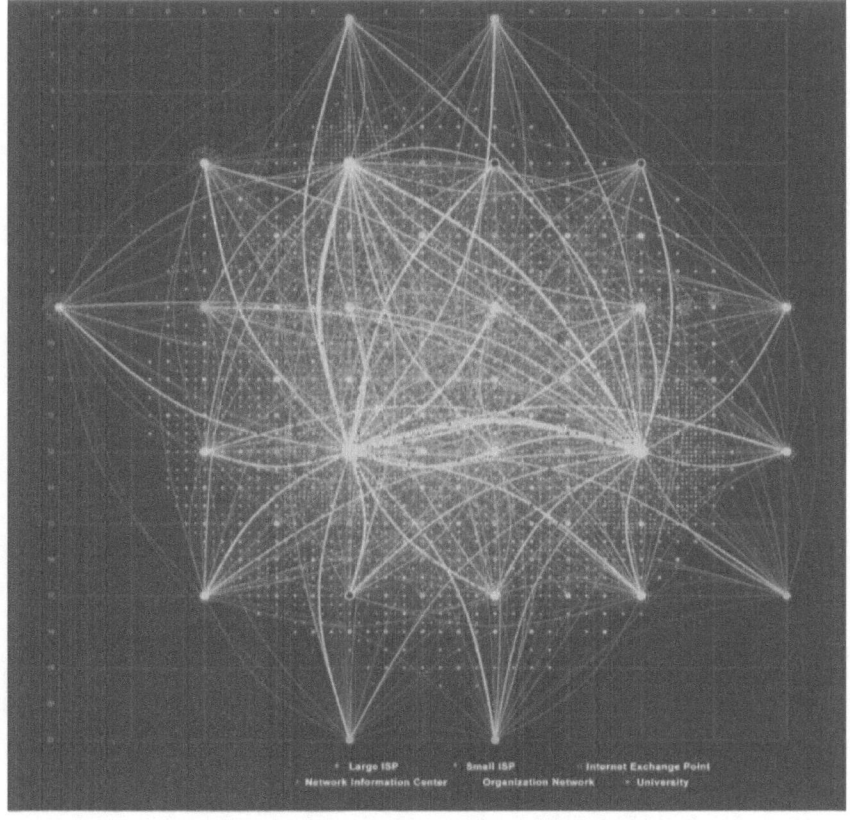

The amount of information being exchanged on the Internet is staggering. By the way, notice that Internet is capitalized. There are local intranets, company's internal internets, etc., but there is only one global Internet. The capital I signifies "The Global Internet"!

But how does this thing work? What are the major processes at work that allow us to see websites? Let's take a look at how a website gets built and put on the Internet. First, we will examine how domain name registration, website hosting, and website design all fit together to create websites. After that, we will show you how the various files that make up a website come together to create website pages.

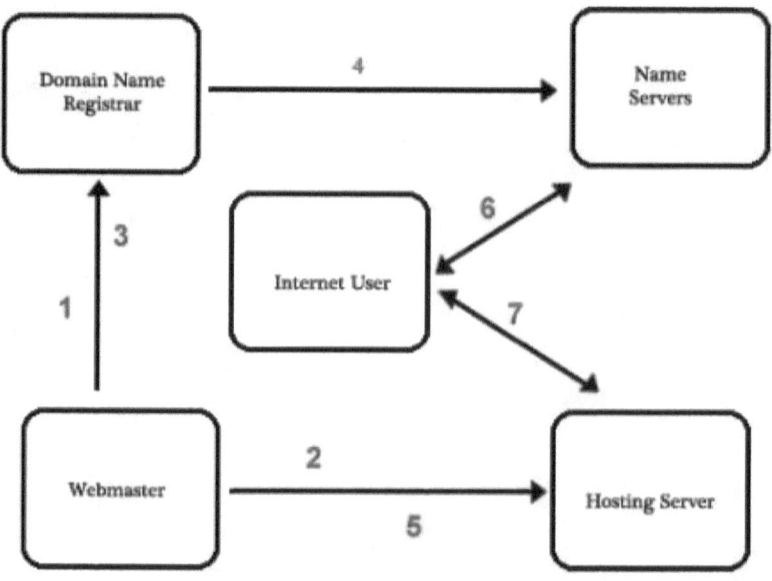

The diagram above shows the various systems that work together to create a website and make it accessible to the public. Each box represents a different computer or server, all interconnected across the Internet. We will first define each of these systems and then discuss each interaction between them.

The Domain Name Registrar is a company capable of registering a domain name, such as abc.com or xyz.org. They assign ownership to that name. The best known registrar is GoDaddy.com.

Name Servers are servers distributed throughout the Internet that can translate a domain name into the physical address (called an IP address) of the server on which that website resides. Your own Internet provider will have a name server, so when you go looking for a website, this server is close by to tell your web browser where to find it.

Internet User is just as it sounds - the end user who is viewing a website.

Webmaster is the person who creates and maintains the website.

Hosting Server is the publicly accessible server on which the website is stored and made available to the public. Hosting servers are not much different than a regular PC; the main difference is that they are connected to the Internet with a very high speed connection, called a T1 line, capable of uploading large amounts of data at very high speed.

Now let's examine how these various pieces interact with one another.

Step 1 - the webmaster logs into a domain name registrar, selects an available domain name, and registers it. The webmaster needs to retain access to the registration account, as he will later come back to the domain registration, and add the name server codes for the hosting server.

Step 2 - the webmaster creates a hosting account for the new website by logging into a hosting service. He obtains login credentials with which he can transfer the website files to the host server using FTP (File Transfer Protocol).

Step 3 - the webmaster now communicates the host server's server codes and by way of them its fixed IP address (its physical address on the Internet) to the domain registration.

Step 4 - the domain registrar distributes the domain name and associated hosting account to all name servers. Updating all the name servers in the world takes several hours, so linking the domain name and the host account is not instantaneous. However, within a few hours, any user computer worldwide will be able to type in the domain name and be directed to the correct hosting server.

Here is our diagram again, just to keep it in view:

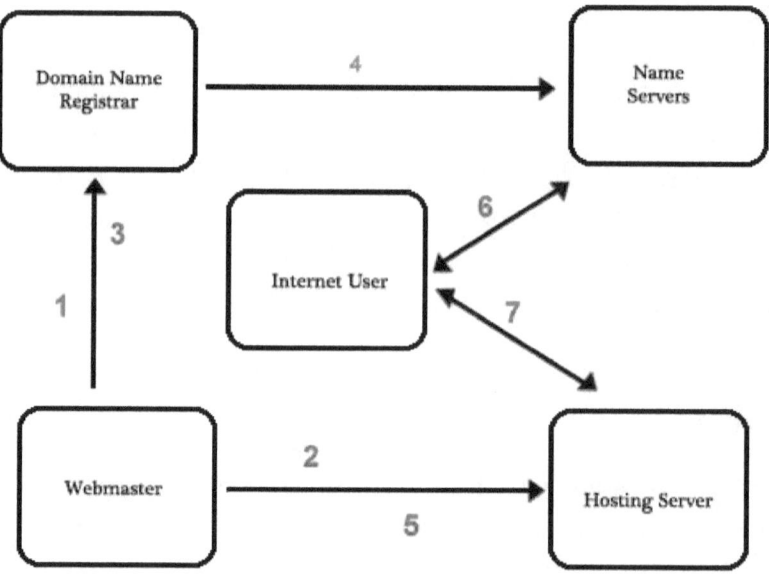

Step 5 - the webmaster can now build the website. The webmaster normally creates the website and all its component files on his own computer, and then uploads it to the hosting server when ready to be viewed by the public.

All of the above basically happens once. We are finally ready for our user to view the website. The next steps happen every time someone views the website.

Step 6 - the user types the domain name into a web browser. The web browser goes to the nearest name server to translate the domain name into the IP address of its host server.

Step 7 - the user's web browser calls the IP address of the host server and requests the main file (known as the index file) for the domain name. The index file will in turn tell the browser what other files are needed to display the home web page.

Now let's look into what actually makes up a website. Some people assume that a web-page must be like a Word document, that is, a single file which is downloaded from a server and displayed on the user's monitor. Actually, it is quite different than that. A single web-page is typically made up of dozens of files, downloaded to the user's computer, and then built up into a web-page "on the fly" by the user's web browser. We'll now look at what those various files are and how they fit together to make a webpage.

The diagram below shows you some of the files that might make up a typical web-page. An HTML file is downloaded first to initiate the process. (HTML stands for HyperText Markup Language. It's the most common programming language used to create web pages.) In the diagram, labelled as item 1, is an HTML file called index.html. The index file is the default first page to be downloaded, so it is typically the file that starts the process of building the Home page.

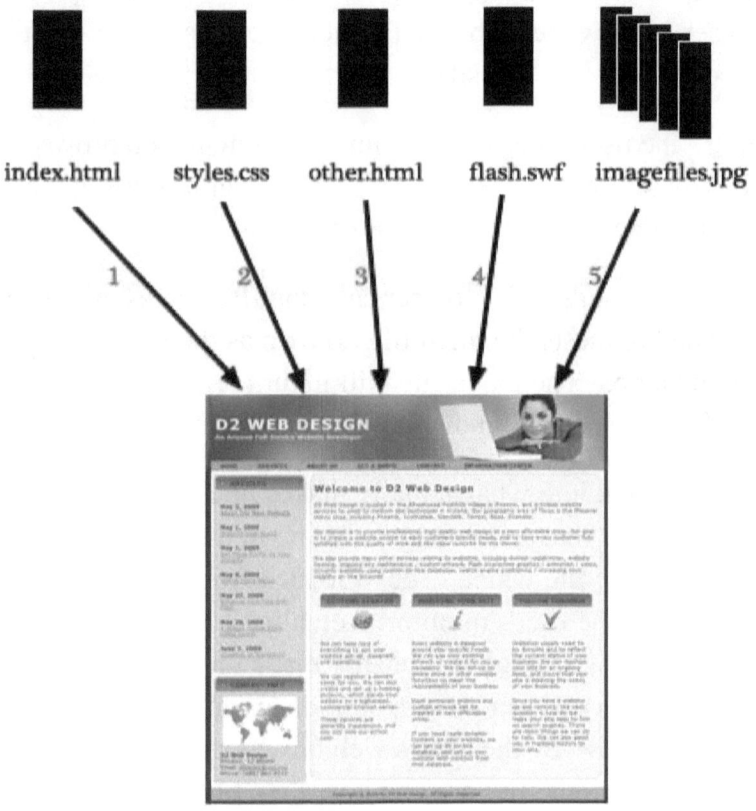

The index file provides all kinds of information about the page that will be assembled. It contains most of the text that will appear on the page. It can contain the header, the menu, the footer information, etc. But it makes numerous references to other resources (files) that will be used to assemble and display the page.

One of the other files referenced by the index file is the style sheet, labeled as item 2 in our diagram. The style sheet specifies the size, location, color, border, font, layout, etc. for each region of the page. By region of the page, we mean things like header, footer, left-hand column, right-hand column, menu area, etc. When a font is specified in a style sheet, it is specified as to size, preferred font, several substitute fonts to be used if the preferred one is not available on the user's computer, and font style (such as bold, underline, italic). The style sheet will also specify how a font will behave if it is part of a link - does it change color when the mouse is over it? Does an underline appear or disappear? Is the link a different color if it has already been visited?

Typically a single style sheet controls every page on the website. This has some great benefits. It guarantees a consistent look across the entire site. It allows the designer to change that look by altering a single file. And it separates the look of the website from the content of the website.

The index file may also call other HTML files as shown in item 3 in our diagram. For example, the webmaster may want to define a complex menu structure in a stand-alone HTML file, and then call it into the index file or other page files. That way, the menu structure is all contained in one place, even though it is repeated and appears on every page.

The index file may also reference a Flash animation, as shown in item 4 in our diagram. Again, each Flash element on a page is a separate file which needs to be downloaded.

Finally, all the images on a web-page are separate files, as shown in item 5 in our diagram. And typically, there are a lot more images on a page than you might expect. While a picture is an obvious image, many other elements of a web-page are also images. For example, anytime you want to completely control the way text looks, perhaps a specific font with no substitutions, or text with a shadow around it, the text has to be placed in an image file. Otherwise, the user's web browser will have the ultimate control of what that text looks like. Similarly, numerous small features on a web-page, such as separators, bullets, rounded corners, etc. are created with images. It is not uncommon for a single web-page to use 40-50 images.

So how does all this come together to create a webpage. Web browsers are so common today that we take them for granted, but they are really very complex pieces of software, capable of reading the instructions from that primary index file, downloading all the auxiliary files necessary to display the page, interpreting all the format instruction in the style sheet file, then assembling the whole thing and displaying it on the user's monitor.

With a slow dial up connection, you can sometimes see this process unfold the first time a web page is viewed, just because of the time involved to get all the files downloaded. With a high-speed connection, the page seems to appear all at once, but a whole lot is going on, right on the user's computer.

Another interesting thing about web browsers is that they anticipate having to use the same files again, either to display another page on the same site or perhaps the same page a few minutes later. All the files are saved for potential reuse - a process referred to as caching. Caching greatly reduces page loading times, as well as total Internet traffic. However, it also means you may not be seeing the most recent changes to a web page.

So now you know at least a little bit about how websites work, and what's inside the black box known as the Internet.

Chapter 14 – Computer Memory

Since we all work with computers today, we can't help but learn a lot of computer terminology. We've all heard about RAM and ROM memory. We know we have a Flash memory card in our cell phone or digital camera. We know there's a hard drive in our computer. We watch movies on DVDs. These are all references to computer memory. But why are there so many different kinds? And what is the purpose of computer memory in the first place? In this chapter, we will attempt to answer those questions, and demystify computer memory.

Let's start with the basic question. Why do computers need memory at all? And what is computer memory?

Computers basically take inputted data, such as numbers or text, process them, and output new data. To do this, a computer needs a place to store the data, at least long enough to operate on it, and it needs a place to store the list of instructions about what it is supposed to do with the data. It needs a place to put the result, at least until it's printed, and perhaps to save it for years.

Computers do everything in binary. The data and the instructions are all encoded in long lists of 1's and 0's, so computer memory is simply a place to store 1's and 0's – usually very large quantities of 1's and 0's are needed to do anything useful.

In the earliest days of computers, memory was accomplished by running wires through little circular magnets. By passing current through the wires, these magnets could be magnetized in one direction or the other to signify a 1 or a 0. Another wire could then be used to retrieve the 1 or 0 stored in that magnet. One magnet was required for each location where a 1 or a 0 was stored, so to get enough memory to do anything useful required filling large rooms with thousands of boards full of wire and magnets.

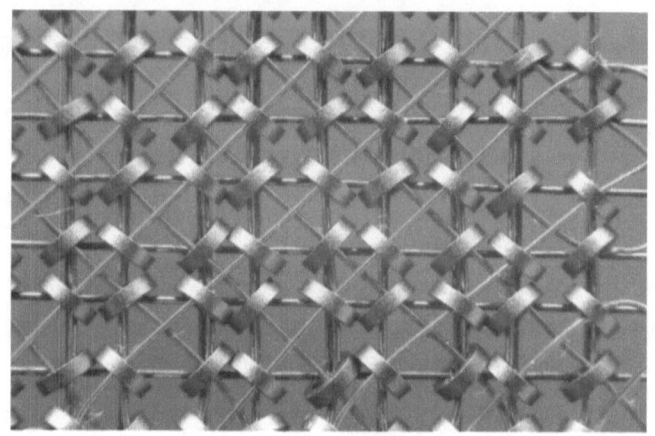

However, aside from being very expensive to build and taking up huge amounts of space, this old "magnetic core" memory had some pretty good properties. It could read or write a lot of 1's and 0's very quickly. It could be accessed randomly, meaning the computer could address any particular location in this vast array of magnets, and read or write a 1 or 0 there. Finally, this memory was what we call "non-volatile", that is, when the computer was shut off and power was removed, the magnets stayed magnetized and remembered whether they stored a 1 or a 0. As we will soon discuss, these are all desirable properties for computer memory.

A big breakthrough occurred in 1970, when Intel introduced the first semiconductor memory chip – the 1103 DRAM chip with 1024 bits of storage on a small, single chip. It replaced a large board with 1000 magnets on it! DRAM stands for dynamic random access memory.

DRAM is still used as the main memory in computers today. That's because DRAM is very dense, meaning we can get the most memory on a chip using DRAM technology. It only requires a tiny, single transistor for each bit stored.

For comparison with the Intel 1103 chip from 1970, today in 2014, we can get 2 GB on a single chip! That's 16,000,000,000 1's or 0's, or 16 million times the capacity of the 1103!

Semiconductor DRAM is pretty fast, and as the name implies, it's random access. But it is volatile; in fact, it's incredible volatile – left to itself, it only can hold its information for a few thousandths of a second, so it has to be constantly refreshed. Refresh means read the data stored before it disappears and then rewrite it in that same location. This constant refresh process requires some special circuitry and means the memory is unavailable to the computer during the refresh process.

DRAM is used for the large working memory of a computer. It's only temporary storage, but it's random access unlike a hard drive. Think about a computer processing an Excel spreadsheet. When you change one number, the computer is able to go around and update all the other numbers on the spreadsheet as a result of your change. This requires being able to quickly move around in memory and implement changes. That's what random access is all about.

The constant need to refresh DRAM slows it down. Because of the refresh process, DRAM is not always available when the

central processor needs data or instructions. So another type of
memory is used by the central processor for data that is changed
quickly and often – it's called static random access memory. It's
more expensive and takes up more space on a chip than DRAM,
but it is much faster and doesn't need to be refreshed. You will
sometimes hear it referred to as the processor's cache memory.
Some static RAM is built into the processor itself and some can
be externally made available to the processor.

Static RAM doesn't not need to refreshed, but it is also volatile.
When power is turned off, static RAM loses its stored
information.   So next we need to look at the various ways to get
non-volatile memory.

The most obvious non-volatile memory that we are all familiar
with is the computer's hard drive. They consist of one or more
magnetic disks rotating at about 120 revolutions per second.

Magnetic heads can record strings of 1's and 0's in tracks on the
magnetic disk. These devices can store incredible amounts of
data. A 1TB (one terabyte) hard drive can store
8,000,000,000,000 1's or 0's!

Hard drives are non-volatile. The data on them lasts for many years. They are very reliable, and the least expensive way to store large amounts of data. But they have two major shortcomings as computer memory goes. They are not random access - data has to be read or written serially in long blocks, so they are not suitable for main memory. And they're also very slow to access compared to DRAM or static RAM.

Another non-volatile memory is CD-ROMs and DVDs. These are similar to hard-drives, but are slower and less expensive. They are optical devices using a laser to reflect light off small pits embedded in the plastic.

Hard drives and CD-ROM/DVD players have one other shortcoming. They are full of mechanical moving parts which wear out and eventually fail. So many other types of non-volatile memory have been developed. These other types are all semiconductor based with no moving parts. There are a number of technologies, and we won't talk about each one, but we will talk about two of the most common.

ROM stands for read-only memory. It's non-volatile, random access, fast, and inexpensive. It's used anywhere that requires a fixed set of computer instructions that aren't going to change. ROMs are hard-wire programmed with those instructions. Smart appliances typically have their program stored in a ROM on the microcontroller. Your personal computer has a small start-up program called the BIOS that starts the process of loading the operating system. It's stored on a ROM.

Flash memory is another type of non-volatile memory. It's fast and inexpensive, can be written and re-written unlike ROMs, but it typically has to be accessed serially, like a hard drive. It is used anywhere that a hard drive might be used. Solid-state hard drives are an example, where a large array of flash memory is used to actually replace the mechanical hard drive in a computer. The memory card in your digital camera, camcorder, and cellphone are also typical applications for flash memory.

So now we have talked about various types of computer memory. We've looked at volatile versus non-volatile. We've looked at random access versus serial access. We've looked at speed versus cost. We've looked at dynamic versus static. Let's try to put it all together and summarize.

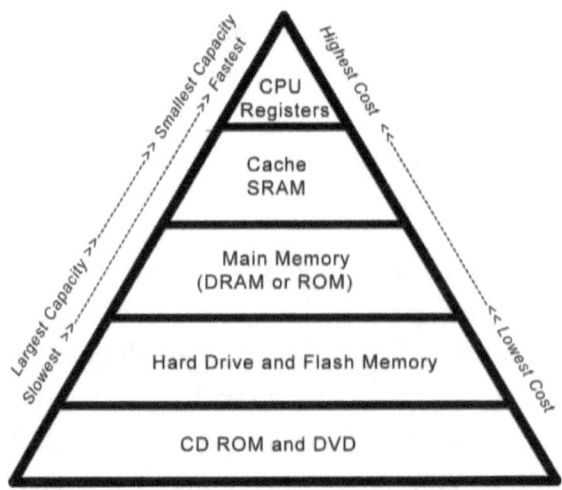

We have many different kinds of computer memory, because the competing technologies are full of trade-offs. We need small amounts of very fast, but expensive memory available to the CPU for speed. We need slower, less expensive memory in large quantities to store our stuff, like pictures, movies, and other data.

So now you know a little bit about all the different kinds of computer memory. They are no longer a black box.

Chapter 15 – Solar Panels

Solar panels are appearing on home rooftops across America, particularly in the southwest where the sun shines 300 days a year. Solar power is growing rapidly, but still only amounts to less than 1% of commercial power produced in the US. The charts below show you where our power came from in 2013.

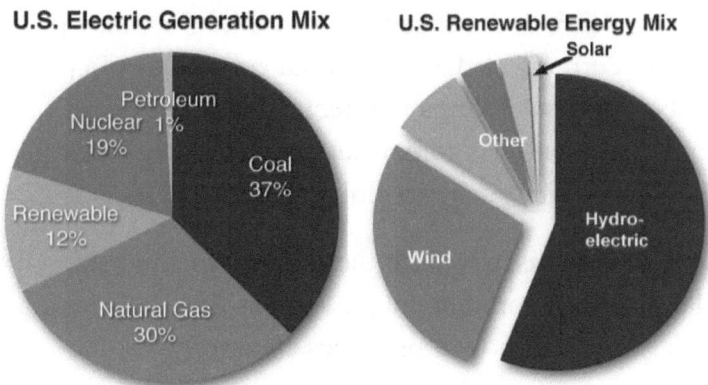

Note that all renewable energy sources only amount to 12% of the electricity we produce, and half of that is hydroelectric power.

The cost of solar panels and solar power in general has been coming down, but cost is still the big issue holding solar back. Otherwise, it would be the ideal power source. Sunlight is plentiful and almost inexhaustible, and solar power does no poluting like coal and gas, and isn't dangerous like nuclear. What's not to like except the cost!

Projections on the growth of solar are "all over the map", but it's currently doubling every 2 years right now, so it's certainly possible for it to be supplying 15-20% of our electricity needs in 20 years.

So now, in this chapter, we'll look at solar panels, and see how they fit into the commercial power system we discussed in Chapter 5. We need to look at solar power in conjunction with commercial power, because in practice, there are times when a home needs more power than solar panels can provide, and there are times when a home can't use all the power the solar panels are producing. So we end up having to deal with the question of how solar panels can be integrated with commercial power.

In our chapter on commercial electric power, we spent a lot of time discussing alternating current and why it's important. Solar panels don't fit into the scheme very well. First they produce DC (direct current), and an individual cell (6" by 6") produces 0.5 volts and about 7 amps in direct sunlight. It doesn't produce power at night, and it's somewhat unreliable, e.g. it stops producing electricity when a cloud passes over it.

We will need to discuss all of these issues to understand the role of solar panels in commercial power, but first, let's look at the solar cell itself and how it produces electricity.

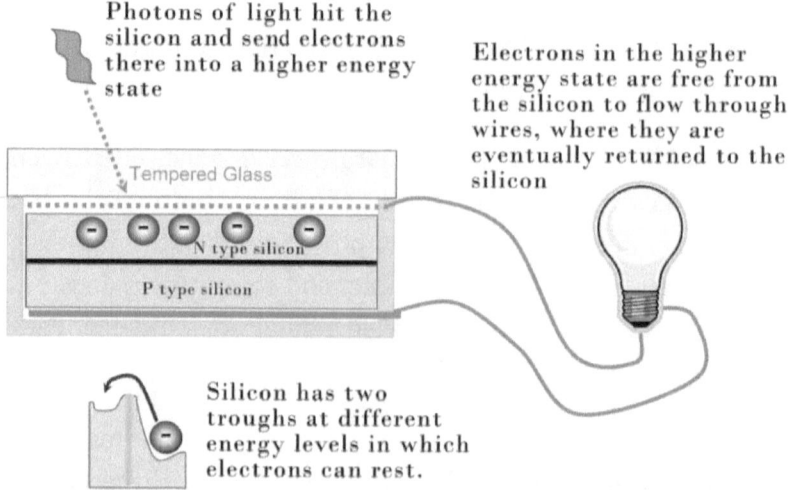

Photons of light hit the silicon and send electrons there into a higher energy state

Electrons in the higher energy state are free from the silicon to flow through wires, where they are eventually returned to the silicon

Tempered Glass

N type silicon

P type silicon

Silicon has two troughs at different energy levels in which electrons can rest.

A solar cell is made of silicon. The top surface is what's called N-type silicon and has an excess of electrons. Below the N-type silicon is P-type silicon, which has a shortage of electrons. When light hits the junction between these two types of silicon, the photons of lights transmit their energy to electrons in the N-silicon, freeing them to flow out of the silicon into wires where they are returned to the silicon on the electron deficient P-type side.

But as previously stated, a solar cell only produces about 0.5 volts, so they have to be wired up in series to form panel segments that produce higher voltages, as shown below.

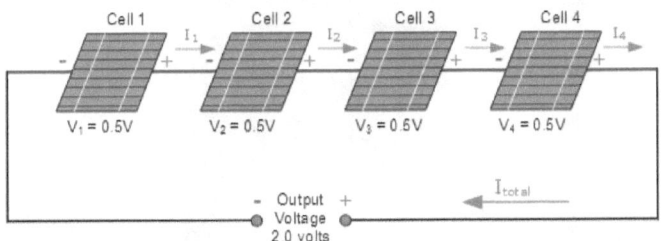

A commercial solar panel section might have 60 cells wired in series so that one section outputs 30 volts at 7 amps. Put 4 sections in a panel and wire them in parallel, and you get a panel that produces 30 volts at 28 amps, which is 840 watts. The drawing below shows a simple combination of series and parallel wiring.

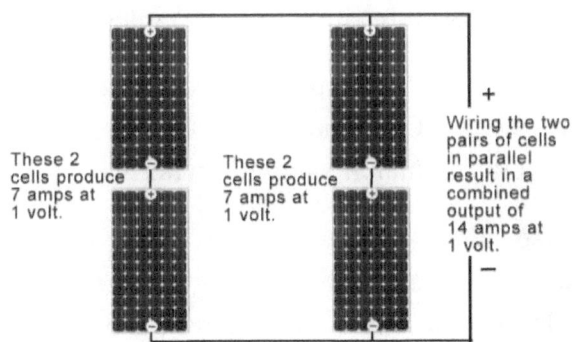

So now, take 5 of our 840 watt panels and put them a roof, and we would have a 4200 watt power system, enough power to power a whole house. However, the 4200 watts is DC (direct current), and its 140 amps at 30 volts. To integrate this power with the commercial power in your home, it has to be converted to 220 volt AC. The device that accomplishes this is called an inverter. One is shown below. Inverters are about 95% efficient at converting DC power to AC.

We need our solar power system to output more or less constant power during the day when the sun is shining. We don't want it shutting down with every passing cloud. So many solar systems include a bank of batteries that provide constant power to the inverter during short interruptions in the solar panels output.

The output of our inverter must be synchronized with the commercial power AC, so the inverter must sense the commercial AC power cycle and match its output with that.

Of course, the solar system won't be working when the sun is down, so even a home with a 4200 watt solar system will need commercial power at night. And there will be many times during the day, when the sun is shining, that the solar power system is putting out much more power than the home is actually using. So homes with solar panels have a special meter that measures how much power the home uses from the power grid, but also measures how much excess power the home puts back into the power grid. This type of measurement system is called "net metering".

So, we are starting to see why solar power is expensive. There are a lot of issues, and a lot of hardware is required besides the panels themselves. The diagram below shows a typical system for integrating a set of solar panels in a home with the AC power grid. We've talked about the panels themselves, the battery back-up, the inverter, and the net metering. However, as you can see from the diagram below, there's a lot of other stuff we haven't even talked about.

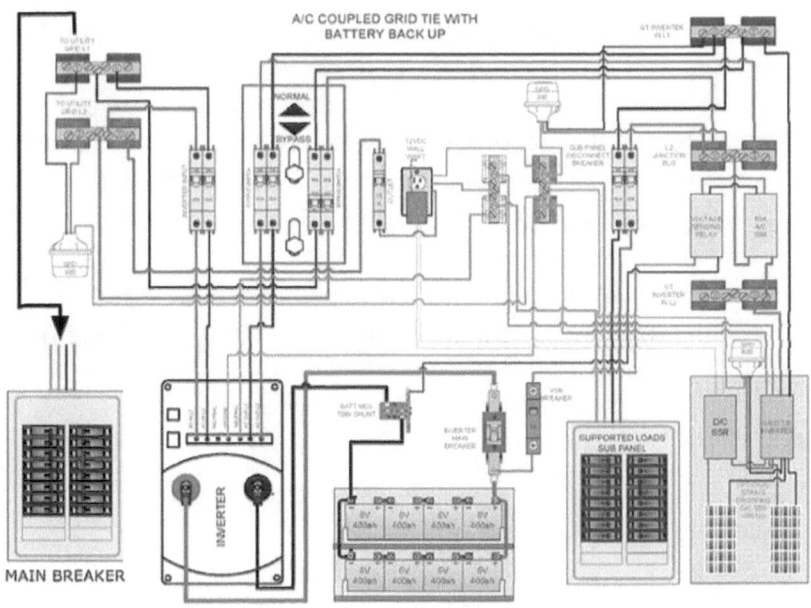

There is another really big issue looming if solar power is going to become a large part of our electric energy supply. While one home can take electricity from the power grid when it needs more and put electricity back into the grid when it makes extra, what happens when solar becomes a large part of our total electicity production? Where will power companies store the excess energy when solar output is at its peak? And where will it get power at night when solar isn't working at all?

What we really need to answer these questions is a huge breakthough in battery technology. Trying to solve this issue with lead-acid batteries like the one in your car, or even big banks of lithion ion batteries is not practical. They are too expensive, take up too much room, and they don't last long enough.

While we are waiting for a big breakthough in batteries, other methods of storing power are being investigated. One is using molten salt to store vast amounts of heat, which could then be used at night to power generators. Another one is using excess energy to break water into hydrogen and oxygen, then use the hydrogen either in fuel cells or as a convensional fuel to power generators.

It's not clear what technologies will emerge to solve the energy storage dilemma, but it is clear that we need one if solar power is going to become a major part of your electricity supply!

So now you know a little about solar panels and how they fit into the commercial power system. They are no longer a black box!

Chapter 16 – A Computer's CPU

At the heart of every computer is a CPU or "central processing unit". It's the guts or inner workings of the computer that actually allow it to do what it does.

Programmers learn to write programs that are executed by the CPU, but what actually happens in the central processor is a mystery even to most programmers, unless they are specifically writing "assembly language" code for the CPU.

My own personal experience with CPUs began when trying to understand the inner workings of Motorola's 6800 micro-processor back in the mid 1970's. A microprocessor is basically a CPU built on a single integrated circuit chip.

The photo above is what the 6800 looked like from the outside. The photo below shows the chip on the inside. It is a complicated circuit made up from 6800 transistor elements.

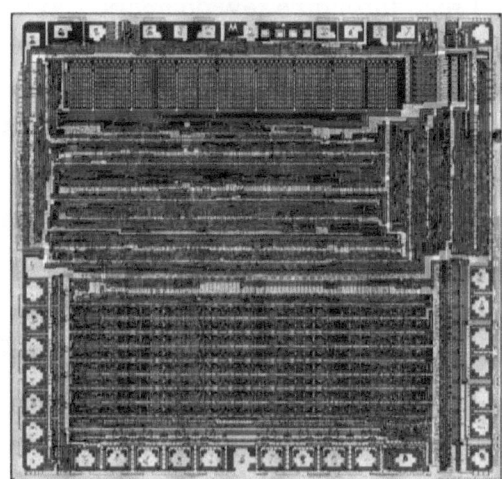

Toward the end of this chapter, after we have defined some of the pieces that make up a CPU, we come back to what all these transistor elements actually do.

Back in 1974, microprocessors were brand new. They made it possible for the first time to build your own computer. I wanted to build one, but first I had to understand how these microprocessors worked. I got a couple of books documenting how to use the 6800. Reading them was a strange experience, as I felt like I was trying to read a book written in an unfamiliar foreign language. Every concept was explained in terms of a bunch of other concepts that I didn't understand.

The first time I read through a few chapters, I'm sure my comprehension was less that 10%. But fortunately, each time I read through it again, I understood a little bit more. Gradually I started to learn this new language and understand what was going on inside this new microprocessor CPU.

In those early days of microprocessors, there were no powerful software tools to help programmers. There were no high-level languages and no software libraries for the 6800. If you were going to get anything done, you had to understand what was going on inside the CPU. So I was forced to learn how the thing worked.

The 6800 was leading-edge technology back in 1974, and putting 6800 transistor elements on a single chip was a major accomplishment back then. However, the modern processors that power today's PCs, MACs, and cellphones are far more powerful than the 6800. They are also much, much more complicated, containing as many as 100 million transistor elements!

But all CPUs basically operate the same, and the stuff they do is really pretty simple. They do it really fast, typically fractions of a microsecond. That means they can perform a thousand operations in a thousandth of a second. But what they do in any one operation is not complicated. And CPUs are also simple in the sense that everything they do is done sequentially, one operation at a time. Soon we will explain what these operations are, but first, we'll need to define some terms.

Let's begin with the concept of a register. It's a place to store one number. Numbers in CPUs are all binary, meaning they are composed of 1's and 0's. A single 1 or 0 is known as a bit, and 8 bits is known as a byte. So, for example, a 16 bit register can store sixteen 1's or 0's, and those sixteen 1's or 0's represent a number between 0 and 65535 (2^{16} possibilities).

CPUs have several different registers. One called the program counter register points to the instruction in memory that is currently being executed. We will talk about program instructions and their execution in a moment. Some other registers called accumulators store numbers for arithmetic processing, such as addition or multiplication. They can also just be a place to temporarily store a number. Other registers such as index registers and stack pointers are places to keep track of something, like where we are in a list we're processing.

Let's now deal with another basic concept – the bus. A bus is a physical pathway, usually between a register in the CPU and a location in memory. It is a certain number of bits wide, meaning it handles that many bits at once. It could be 8 bits wide and handle numbers between 0 and 255 (2^8 possibilities). Or it could be 64 bits wide and handle numbers between 0 and 18,446,744,073,709,551,615 (2^{64} possibilities). One of these buses is an address bus; it's used to connect the CPU to a specific location in memory. Another bus is a data bus; it is used by the CPU to read data from or write data into a specific address in memory.

Now let's talk about the processor's instruction set. It's a list of operations the processor is capable of carrying out. These operations are all simple stuff. The power comes not from any one instruction, but from the ways they can be ingeniously combined to accomplish complex tasks. For example, one instruction might say "load accumulator B with the data stored at address 23454321". Another instruction might be "increase the content of accumulator B by 5". And another could be "put the content of accumulator B into the data stored at the memory address currently pointed to by the index register".

To see how you might get something useful done with instructions like this, let's suppose we want to output a line of text to a printer. The text is stored in memory starting at address 345678 and is 28 characters long. A character might be one letter, a period, or a space. Each character is stored in one byte of memory (8 bits). Each byte has its own address in memory, so if the first character was stored at address 345678, then the second character would be stored at 345679, etc. Let's also assume that there is an 8 bit register called print-out. Data sent to the print-out register is automatically passed on to the printer. Let's see the kind of CPU instructions that would cause our text to print.

1. Load index register with starting address of the text: 345678.
2. Load accumulator A with the number of characters: 28.
3. Load accumulator B with content pointed to by the index

register: the first character in our text

4. Store the content of accumulator B in the print-out register.

5. Increment index register by 1.

6. Decrement accumulator A by 1.

7. If accumulator A content is greater than 0, go back to step 3 and repeat.

This code would loop around 28 times – each time reading a character and sending it to the printer. The CPU would actually have to execute 142 operations in order to print our text.

The programming language of the CPU is usually called assembly language. A program called an assembler is used to write assembly language. An example of assembler code for the 6800 microprocessor is below. (It's just a sample to see what assembly language looks like – not our code to print a line of text.)

```
MONITOR FOR 6800 1.4          9-14-80  TSC ASSEMBLER  PAGE    2

C000                    ORG     ROM+$0000 BEGIN MONITOR
C000 8E 00 70  START    LDS     #STACK
C003 86 13     INITA    LDA A   #RESETA   RESET ACIA
C005 B7 80 04           STA A   ACIA
C008 86 11              LDA A   #CTLREG   SET 8 BITS AND 2 STOP
C00A B7 80 04           STA A   ACIA
C00D 7E C0 F1           JMP     SIGNON    GO TO START OF MONITOR
C010 B6 80 04  INCH     LDA A   ACIA      GET STATUS
C013 47                 ASR A             SHIFT RDRF FLAG INTO CARRY
C014 24 FA              BCC     INCH      RECIEVE NOT READY
C016 B6 80 05           LDA A   ACIA+1    GET CHAR
C019 84 7F              AND A   #$7F      MASK PARITY
C01B 7E C0 79           JMP     OUTCH     ECHO & RTS
C01E 8D F0     INHEX    BSR     INCH      GET A CHAR
C020 81 30              CMP A   #'0       ZERO
C022 2B 11              BMI     HEXERR    NOT HEX
C024 81 39              CMP A   #'9       NINE
C026 2F 0A              BLE     HEXRTS    GOOD HEX
C028 81 41              CMP A   #'A
C02A 2B 09              BMI     HEXERR    NOT HEX
C02C 81 46              CMP A   #'F
C02E 2E 05              BGT     HEXERR
C030 80 07              SUB A   #7        FIX A-F
C032 84 0F     HEXRTS   AND A   #$0F      CONVERT ASCII TO DIGIT
C034 39                 RTS

C035 7E C0 AF  HEXERR   JMP     CTRL      RETURN TO CONTROL LOOP
```

Programmers today typically write software with high-level language compilers like C++, Python, or Java. They are powerful tools that make it much easier to program complex tasks. The compiler processes the high-level language code and translates it into machine code. The programmer doesn't need to know assembly language or what the CPU is actually doing. In the high-level language, he would just write the code:

Print "This is a sentence to print."

The compiler would then generate the machine code required. (Note our sentence above has the same 28 characters we used in our earlier example. Spaces and the period are themselves characters.)

Our assembler or our high-level language ultimately creates machine code – the stuff that is actually executed by the CPU. What does machine code look like? An example is below.

```
3F20    3F 20 3F F6 0C 8E A0 60 CE 3F B7 A6 00 C6 2E 11
3F30    27 06 BD E1 D1 08 20 F3 86 0D BD E1 D1 86 0A BD
3F40    E1 D1 86 00 BD E1 D1 BD E1 D1 CE 3F CF A6 00 11
3F50    27 06 BD E1 D1 08 20 F5 8D 51 FF 3F 20 BD E0 CC
3F60    CE 3F E3 A6 00 C6 2E 11 27 06 BD E1 D1 08 20 F3
3F70    8D 39 08 FF 3F 22 FE 3F 20 86 0D BD E1 D1 86 0A
3F80    BD E1 D1 86 11 B7 3F 24 FF 3F 20 CE 3F 20 BD E0
3F90    C8 FE 3F 20 7A 3F 24 27 E0 BD E0 CC A6 00 BD E0
3FA0    BF BC 3F 22 27 02 20 EC 7E E0 E3 BD E0 CC 86 3F
3FB0    BD E1 D1 BD E0 47 39 48 45 58 41 44 45 43 49 4D
3FC0    41 4C 20 4D 45 4D 4F 52 59 20 44 55 4D 50 2E 46
3FD0    49 52 53 54 20 42 59 54 45 20 54 4F 20 50 52 49
3FE0    4E 54 2E 4C 41 53 54 20 42 59 54 45 20 54 4F 20
3FF0    50 52 49 4E 54 2E
```

Machine code is basically a list of numbers. They are show in the example above in hexadecimal (base 16) format. (Why hexadecimal? It allows us to express an 8 bit or 1 byte number as two digits - you can't do that with decimal, as there are 256 possible numbers in a byte.) Some of the numbers are actual instructions, while others are data to load, memory addresses, etc. As the program counter goes through this list, the processor is able to decode it into actual instructions and execute them. For example, 86 is an instruction for the 6800 that says "load accumulator A with the byte that immediately follows the instruction itself". If the next byte is 3F, which is hexadecimal for 63, then 63 would be loaded into accumulator A by this instruction. In general, machine code is specific to a particular CPU, so it only works with the specific CPU it was written for.

Now the next thing you might want to ask is "How does the CPU actually do any of this?" Everything is basically hard-wired into the chip. The registers, the busses, the logic required to execute the instruction set, the arithmetic processes, etc. are all hard-wired circuitry built into the integrated circuit. Those 6800 transistor elements are wired together to make the registers and other physical hardware needed to make the CPU work.

So now you know a little bit about the CPU at the heart of a computer. I won't say it's no longer a black box, but hopefully, it's not quite as mysterious as it was!

Chapter 17 – Digital Cameras

Today digital cameras have almost completely replaced conventional film cameras, but how do they work? We'll take a look in this chapter.

Let's begin with how cameras work in general, and then go into some of the specifics about how a digital camera works.

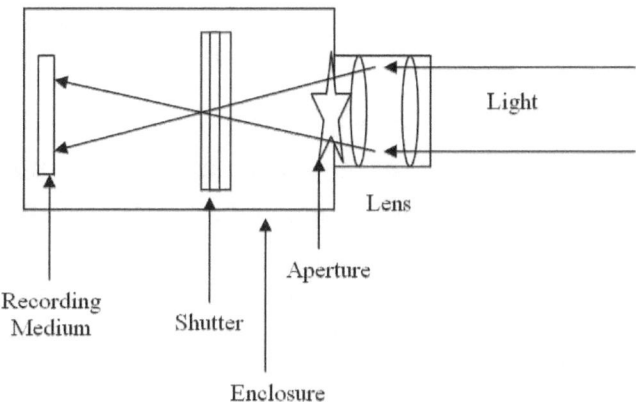

The diagram above shows the four main parts of any camera.

The lens is the thing that focuses the image on a recording medium like film.

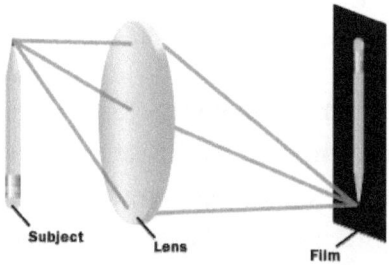

Lenses create an image of our subject on the recording medium. This image is inverted in the process so that it appears upside down. By changing focal length of the lens (the distance between the lens and the film or recording medium), we can make it a telephoto lens, or a wide angle lens. Some lenses have a fixed focal length, but zoom lenses allow the user to adjust the focal length at will.

The aperture is the hole that lets in the light. It can be adjusted in size to control the amount of light entering the camera.

The shutter is the screen covering the recording medium until you are ready to take your picture. The button you press to take the picture is called the shutter release. It causes the shutter to move out of the way for a brief period which can usually be controlled by the camera user.

The aperture and shutter speed together control the amount of light that reaches the recording medium. Regardless of what the recording medium is, getting the proper amount of light to it is critical to getting a good photograph. Using a short shutter time and a fairly open aperture allows you to stop motion, and get clear images of fast moving subjects. On the other hand, using a long shutter time and closing the aperture as much as possible gives you a wider focal range and a generally sharper image.

Some cameras give the user complete control of aperture and shutter speed, while others are completely automatic. Still others ask you what kind of picture you're taking, and then determine what settings are appropriate.

The recording medium is what records the image so it can be viewed later. For a conventional camera, it's a piece of film. With digital cameras, it's a large number of light sensitive sensors, which is often referred as a sensor array.

Now let's talk about focus. Some cameras using a small aperture don't bother with focus at all. Others ask the user to focus the image manually. Still others have built-in auto-focus systems that always guarantee a sharp image.

So, cameras come in many different configurations, whether they use film or digital recording medium.

Now let's talk about the digital camera. With digital cameras, the recording medium is usually a CCD sensor array. Understanding the digital camera requires an understanding of this sensor.

CCD stands for charged couple device. When we talk about how many mega-pixels a camera has, we are talking about how many tiny little regions on this device can store an electric charge. The electric charge these devices store is directly proportional to the light that has hit them, as they convert photons of light into electric charges. The term charge coupling comes from how the image is retrieved from the array. The device is capable of shifting the charges across the array, so that they can all be collected at one edge. Then they can be sent out serially (meaning one after another) to hardware that converts the charge into a digital number which can be stored in memory. It's impressive that 5 or 10 million pixels in a digital image can go through this whole process quickly enough that we perceive the picture to have been taken instantaneously.

A CCD sensor array can only measure the total amount of light that hits a given pixel in the array. It doesn't distinguish between colors, so how are color images captured?

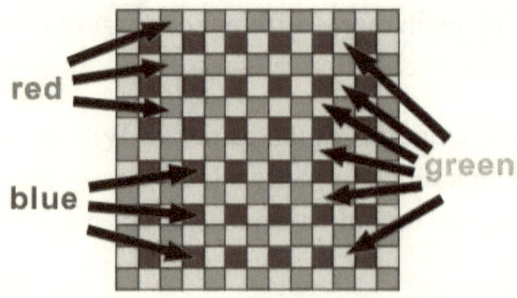

A color filter, known as a Bayer filter, is laid down on top of the array. Every pixel sees only one color as a result of this filter. As you can see, there are twice as many green pixels as there are red or blue in a Bayer filter. That's because our eyes are more sensitive to green than they are to blue or red, so this mix of colors most closely matches our natural perception of color.

What happens next to actually get a color image? In the process of digitizing the image, each pixel is evaluated for color. That's done not just by looking at its own color, but looking at the intensity of color from all the pixels surrounding it. By combining this data, we can create an estimate of the true color for each pixel.

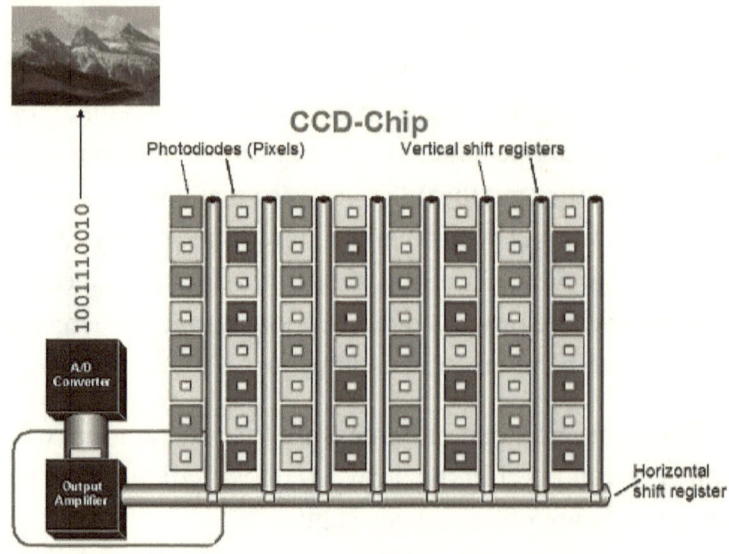

The image on the previous page illustrates the whole process. Again, that this can happen for millions of pixels in such a short period of time is quite amazing!

Now let's look at the rest of a digital camera. There's a lot to talk about. We won't cover everything, but we will look at a few of the many things a digital camera must do.

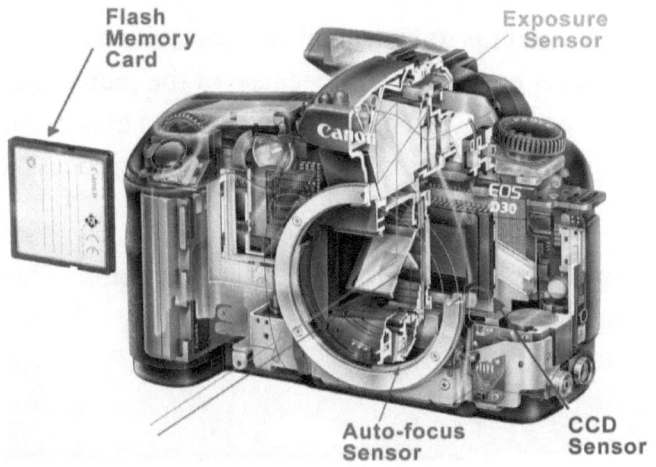

First, you might want to notice in the cutaway view above, that the CCD sensor is sitting directly behind the lens. And, it's surrounded by a bunch of electronics.

There are a couple of other sensors in the camera besides the CCD sensor. An exposure sensor must measure the average light level of the scene being photographed and then, based on that, set the aperture and shutter speed appropriately. This process is initiated on most digital cameras by pressing the shutter release button down part way. The exposure sensor is also used to determine whether the flash is turned on or off.

Another sensor is the auto focus sensor. It moves the lens in or out until a sharply focused image is detected. Many digital cameras have software to detect faces in the image and can actually set the focus to make sure faces specifically are in focus.

The main memory for actually storing the image is a flash memory card. The digital images are stored as numbers on this card and are usually compressed in size using a compression technique such as jpeg format. Twenty years ago when digital cameras first became popular, the limited number of pictures you could store in memory was a big limitation. Today, the typical flash memory card can store thousands of images.

Most digital cameras today can also record video at 30 frames per second. This is possible again because of the incredible speed of the electronics in modern digital cameras, and also because of some fantastic compression techniques used to process and store video.

So now you know a little bit about digital cameras. They are no longer just black boxes.

Chapter 18 – Aircraft Black Box

In a book about black boxes, we can't ignore the aircraft black box – the flight data and cockpit voice recorders in every commercial aircraft!

By coincidence, I am writing this in April 2014, as the whole world is watching the search for Malaysia Airlines Flight 370, which mysteriously disappeared without a trace over a month ago in a remote part of the Indian Ocean. If the black boxes from this aircraft are ever recovered, we will finally start to learn what went wrong on Flight 370!

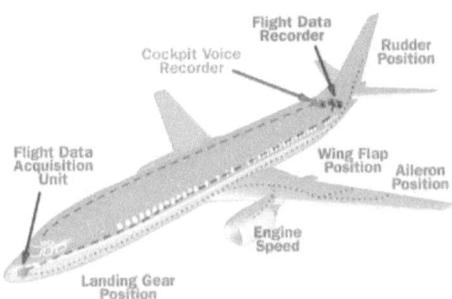

The devices themselves sit in the very back of the plane, where they are most likely to survive a severe crash.

As you may already know, aircraft black boxes are not black. They might have started out black, but the color was quickly switched to the brightest, easiest to find, colors possible, and today, the industry has standardized on orange.

Let's talk about the cockpit voice recorder first. It records the sound from 4 sources – the pilot's microphone, the co-pilot's microphone, the microphone of a third person (flight engineer) in the cockpit, if there is one, and a microphone in the middle of the cockpit which can pick up other sounds, like the click of a switch. In the older versions of the cockpit voice recorder, which used magnetic tape, it recorded the sounds in the cockpit for 30 minutes in a continuous loop, so that only the last 30 minutes were available. Today's modern version of the device digitally records the sounds in solid-state digital memory (Flash memory, if you read the chapter on computer memory) and can record the last 2 hours of sounds in the cockpit.

Now let's discuss the flight data recorder. Again, it started out using magnetic tape, but the modern version uses solid-state memory and can store 25 hours of flight data. Here is small sampling of the type of data recorded:

Time	Control-column position
Pressure altitude	Rudder-pedal position
Airspeed	Control-wheel position
Vertical acceleration	Horizontal stabilizer
Magnetic heading	Fuel flow
Auto pilot settings	Cabin pressure

In practice, modern flight data recorders now monitor and record thousands of parameters, and the accumulated information recorded over 25 hours can amount to millions of megabytes of data (A terabyte is the term for a million megabytes).

For both the cockpit voice recorder and the flight data recorder, the solid state memory arrays are the heart of the device and the thing that must survive after a high-impact crash. For this reason, they are encased and insulated with three layers of material. First, starting on the inside, the memory array is enclosed in an aluminum housing. Next, about 1 inch of dry-silica material provides high-temperature insulation. On the outside, an armor casing made of stainless-steel or titanium encloses the whole thing.

Black boxes undergo very rigorous testing to insure they will survive a crash. They must be able to survive impact, crushing, fire, salt water, deep sea submersion, etc.

Another requirement for black boxes is that they can be located after a crash in the ocean. They are equipped with a device called a pinger which emits ultrasonic pulses that can be detected by sonar or other underwater search equipment. The pinger sends out a pulse once every second. The batteries in the device can power it for about one month.

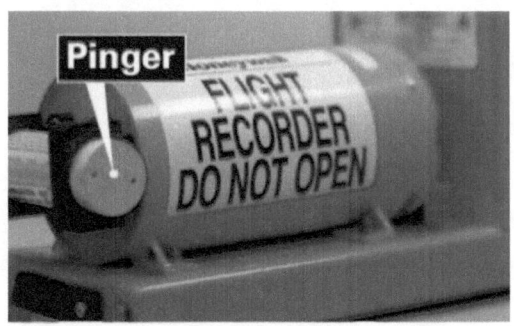

It might be interesting to speculate about the future of aircraft black boxes. They may record cockpit video in the future. That would give investigators additional information, although pilots seem to be resisting.

The other thing that is changing is that aircraft today, though not always in contact with ground control, can be in continuous communication with satellites. All the data going to black boxes could be transmitted to satellites. If satellites were relaying the data to ground stations, and it was being recorded there, there would no longer be a need for black boxes. The big advantage would be that we would no longer have to go looking for them and hoping to find them intact.

So now you know a little bit about aircraft black boxes. I'm tempted to say they are no longer black boxes, but I guess they will continue to be called black boxes!

Final Thoughts

I've always been fascinated with how stuff works. And I also tend to be someone who prides himself in knowing at least a little bit about a lot of different technologies. Of course the downside of knowing a little bit about a number of different subjects is that you aren't an expert of any of them. However, it has given me the ability to write a book like this!

We've looked at a lot of common technologies that you likely use every day. We have tried to demystify them, and tell you a little bit about how they work.

You obviously don't need to know how these things work to enjoy them and make good use of them. However, as you get more familiar with how things work, you can better understand the trends in technology. And you will also start to see certain technologies appear over and over again in a variety of different products.

www.ingramcontent.com/pod-product-compliance
Lightning Source LLC
Chambersburg PA
CBHW022009170526
45157CB00003B/1210